Springer Proceedings in Physics

Volume 196

The series Springer Proceedings in Physics, founded in 1984, is devoted to timely reports of state-of-the-art developments in physics and related sciences. Typically based on material presented at conferences, workshops and similar scientific meetings, volumes published in this series will constitute a comprehensive up-to-date source of reference on a field or subfield of relevance in contemporary physics. Proposals must include the following:

- name, place and date of the scientific meeting
- a link to the committees (local organization, international advisors etc.)
- scientific description of the meeting
- list of invited/plenary speakers
- an estimate of the planned proceedings book parameters (number of pages/ articles, requested number of bulk copies, submission deadline).

More information about this series at http://www.springer.com/series/361

Ramis Örlü · Alessandro Talamelli
Martin Oberlack · Joachim Peinke
Editors

Progress in Turbulence VII

Proceedings of the iTi Conference in Turbulence 2016

Springer

Editors
Ramis Örlü
Linné FLOW Centre
KTH Mechanics
Stockholm
Sweden

Martin Oberlack
Department of Mechanical Engineering
TU Darmstadt
Darmstadt
Germany

Alessandro Talamelli
Department of Industrial Engineering, Alma Mater Studiorum
Università di Bologna
Forlì
Italy

Joachim Peinke
Institute of Physics, AG TWiSt—Turbulenz, Wind energy and Stochastics, ForWind—Center for Wind Energy Research
Carl-von-Ossietzky University Oldenburg
Oldenburg
Germany

ISSN 0930-8989 ISSN 1867-4941 (electronic)
Springer Proceedings in Physics
ISBN 978-3-319-57933-7 ISBN 978-3-319-57934-4 (eBook)
DOI 10.1007/978-3-319-57934-4

Library of Congress Control Number: 2017939896

Printed on acid-free paper

This Springer imprint is published by Springer Nature
The registered company is Springer International Publishing AG
The registered company address is: Gewerbestrasse 11, 6330 Cham, Switzerland

Preface

The iTi has become an established biannual conference on turbulence research taking place in the years between the ETC—(European Turbulence Conference) and TSFP—(Turbulence and Shear Flow Phenomena) conferences. With 80 to 100 participants, the iTi conference places value on the discussions and personal contacts in the location of the beautiful town of Bertinoro in Northern Italy close to Bologna. It continues a tradition that has been started in Bad Zwischenahn/ Germany with the first edition of the conference in 2003. The size of the conference allows to have no parallel sessions and gives time to special topics to be stressed. The content-related focus areas of the conference are the interdisciplinary aspects of turbulence, defining the abbreviation iTi—interdisciplinary Turbulence initiative. iTi attracts scientist from the engineering, physics, and mathematics communities.

It has been a tradition of the iTi to organize a one-day workshop before the iTi conference on a distinct theme out of the wide spectrum of turbulence research. The present workshop was on*High Reynolds number turbulent flows—A large-scale infrastructure perspective*. The 7th iTi in 2016 conference hosted 90 scientists from 15 different countries. In total, there were 78 contributions, from which 50 were presented as talks, with six invited talks, covering a wide range of aspects of current turbulence research. Advances in the basics of understanding and modeling turbulence were addressed as well as practical implications such as the control of turbulence.

The content of the 7th iTi conference is documented in this volume comprising 35 contributions. All contributions were thoroughly reviewed by external reviewers, to whom we want to express our thanks for their valuable and important contribution. Both the workshop and conference were sponsored by the European High-performance Infrastructures in Turbulence (EuHIT). EuHIT is an international scientific mobility programme for researchers engaged in turbulence research (www.euhit.org).

Based on the successful previous conferences, we will continue with this initiative for subsequent years with the 8th iTi Conference planned for September 2018.

Stockholm, Sweden — Ramis Örlü
Forlì, Italy — Alessandro Talamelli
Darmstadt, Germany — Martin Oberlack
Oldenburg, Germany — Joachim Peinke
2017

Contents

Contributors

Abouelmagd Abdelsamie University of Magdeburg “Otto von Guericke”, Universitätsplatz 2, Magdeburg, Germany

P. Henrik Alfredsson Linné Flow Centre, KTH Royal Institute of Technology, Stockholm, Sweden

Diego Angeli DISMI, University of Modena and Reggio Emilia, Reggio Emilia, Italy

Jérémy Basley LHEEA, Ecole Centrale Nantes (CNRS UMR 6598), Nantes Cedex, France

Patrick Bechlars University of Melbourne, Parkville, Australia

Gabriele Bellani Department of Industrial Engineering, Università di Bologna, Forli, Italy

Amitabh Bhattacharya Indian Institute of Technology Bombay, Mumbai, India

Alexandra Bobke Linné FLOW Centre, KTH Mechanics, Stockholm, Sweden; Swedish e-Science Research Centre (SeRC), Stockholm, Sweden

Massimiliano Breda Imperial College London, London, UK

Preben Buchhave Intarsia Optics, Birkerød, Denmark

Oliver R.H. Buxton Imperial College London, London, UK

Jacopo Canton Linné FLOW Centre, KTH Mechanics, Stockholm, Sweden

Cheng Chi University of Magdeburg “Otto von Guericke”, Universitätsplatz 2, Magdeburg, Germany

Andrea Cimarelli DISMI, Università degli Studi di Modena e Reggio Emilia, Reggio Emilia, Italy

Andrea Crivellini Università Politecnica delle Marche, Ancona, AN, Italy

Elisabetta De Angelis School of Engineering, Cardiff University, Cardiff, UK

Sébastien Deck Onera The French Aerospace Lab, Meudon, France

Stefano Discetti Aerospace Engineering Group, Universidad Carlos III de Madrid, Leganés, Spain

Artur Dróżdż Czestochowa University of Technology, Czestochowa, Poland

Christoph Egbers Department of Aerodynamics and Fluid Mechanics, Brandenburg University of Technology, Cottbus, Germany

Peter W. Egolf University of Applied Sciences of Western Switzerland, Yverdon, Switzerland

Witold Elsner Czestochowa University of Technology, Czestochowa, Poland

Bengt E.G. Fallenius Linné Flow Centre, KTH Royal Institute of Technology, Stockholm, Sweden

Marco Ferro Linné Flow Centre, KTH Royal Institute of Technology, Stockholm, Sweden

Tommaso Fiorini Department of Industrial Engineering, Università di Bologna, Forli, Italy

Matteo Franciolini Università Politecnica delle Marche, Ancona, AN, Italy

Jens H.M. Fransson Linné Flow Centre, KTH Royal Institute of Technology, Stockholm, Sweden

Bettina Frohnapfel Karlsruhe Institute of Technology, Karlsruhe, Germany

Noriyuki Furuichi National Institute of Advanced Industrial Science and Technology (AIST), National Metrology Institute of Japan (NMIJ), Tsukuba, Japan

Davide Gatti Karlsruhe Institute of Technology, Karlsruhe, Germany

Gerd Gülker ForWind, Institute of Physics, University of Oldenburg, Oldenburg, Germany

Yosuke Hasegawa University of Tokyo, Tokyo, Japan

Arman Hemmati Department of Mechanical and Aerospace Engineering, Princeton University, Princeton, NJ, USA; University of Calgary, Calgary, AB, Canada

Azur Hodžić Technical University of Denmark, Kgs. Lyngby, Denmark

Michael Hölling ForWind, Institute of Physics, University of Oldenburg, Oldenburg, Germany

Shun Horii Mechanical Systems Engineering, Shinshu University, Nagano, Japan

Sergio Hoyas Instituto de Matemática Pura y Aplicada, Universitat Politècnica de València, Valencia, Spain

Kolumban Hutter Swiss Federal Institute of Technology, ETH, Zurich, Switzerland

Lorenz Hufnagel Linné FLOW Centre, KTH Mechanics, Stockholm, Sweden

Andrea Ianiro Aerospace Engineering Group, Universidad Carlos III de Madrid, Leganés, Spain

Gábor Janiga University of Magdeburg "Otto von Guericke", Universitätsplatz 2, Magdeburg, Germany

Takuya Kawata Linné Flow Centre, KTH Royal Institute of Technology, Stockholm, Sweden

Stefanie Kraheberger Chair of Fluid Dynamics, TU Darmstadt, Darmstadt, Germany; Graduate School of Excellence Computational Engineering (GSCE), TU Darmstadt, Darmstadt, Germany

Cristian C. Lalescu Max Planck Institute for Dynamics and Self-Organization, Göttingen, Germany

Tae-Woo Lee Mechanical and Aerospace Engineering, SEMTE, Arizona State University, Tempe, AZ, USA

Adriano Leonforte DISMI, University of Modena and Reggio Emilia, Reggio Emilia, Italy

Jun-De Li College of Engineering and Science, Victoria University, Melbourne, Australia; Institute for Turbulence-Noise-Vibration Interactions and Control, Shenzhen Graduate School, Harbin Institute of Technology, Shenzhen, China

Laura J. Lukassen Max Planck Institute for Dynamics and Self-Organization, Göttingen, Germany

Oana Marin MCS, Argonne National Laboratory, Lemont, USA

Robert J. Martinuzzi Department of Mechanical and Manufacturing Engineering, University of Calgary, Calgary, AB, Canada

Masaharu Matsubara Mechanical Systems Engineering, Shinshu University, Nagano, Japan

Elia Merzari MCS, Argonne National Laboratory, Lemont, USA

Makoto Miyazaki Mechanical Systems Engineering, Shinshu University, Nagano, Japan

Ingrid Neunaber ForWind, Institute of Physics, University of Oldenburg, Oldenburg, Germany

Alessandra Nigro Università della Calabria, Rende, CS, Italy

Azad Noorani Linné FLOW Centre, KTH Mechanics, Stockholm, Sweden

Martin Oberlack Chair of Fluid Dynamics, TU Darmstadt, Darmstadt, Germany; Graduate School of Excellence Computational Engineering (GSCE), TU Darmstadt, Darmstadt, Germany

Emir Öngüner Department of Aerodynamics and Fluid Mechanics, Brandenburg University of Technology, Cottbus, Germany

Ramis Örlü Linné FLOW Centre, KTH Mechanics, Stockholm, Sweden

Chandra Shekhar Pant Indian Institute of Technology Bombay, Mumbai, India

Jung Eun Park Mechanical and Aerospace Engineering, SEMTE, Arizona State University, Tempe, AZ, USA

Joachim Peinke ForWind, Institute of Physics, University of Oldenburg, Oldenburg, Germany

Laurent Perret LHEEA, Ecole Centrale Nantes (CNRS UMR 6598), Nantes Cedex, France

Maurizio Quadrio Politecnico di Milano, Milano, Italy

Ronald A. Remmerswaal University of Groningen, Groningen, The Netherlands

Nicolas Renard Onera The French Aerospace Lab, Meudon, France

Yoshiyuki Sagawa Mechanical Systems Engineering, Shinshu University, Nagano, Japan

Richard D. Sandberg University of Melbourne, Parkville, Australia

Carlos Sanmiguel Vila Aerospace Engineering Group, Universidad Carlos III de Madrid, Leganés, Spain

Antonio Segalini Linné FLOW Centre, KTH Mechanics, Stockholm, Sweden

Philipp Schlatter Linné FLOW Centre, KTH Mechanics, Stockholm, Sweden; Swedish e-Science Research Centre (SeRC), Stockholm, Sweden

Jannik Schottler ForWind, Institute of Physics, University of Oldenburg, Oldenburg, Germany

Amir Shahirpour Department of Aerodynamics and Fluid Mechanics, Brandenburg University of Technology, Cottbus, Germany

Olga Shishkina Max Planck Institute for Dynamics and Self-Organization, Göttingen, Germany

Maurits H. Silvis University of Groningen, Groningen, The Netherlands

Alessandro Talamelli Department of Industrial Engineering, Università di Bologna, Forli, Italy

Yoshiya Terao National Institute of Advanced Industrial Science and Technology (AIST), National Metrology Institute of Japan (NMIJ), Tsukuba, Japan

Dominique Thévenin University of Magdeburg "Otto von Guericke", Universitätsplatz 2, Magdeburg, Germany

Riccardo Togni DIN, Università di Bologna, Forlì, Italy

Dominik Traphan ForWind, Institute of Physics, University of Oldenburg, Oldenburg, Germany

Yoshiyuki Tsuji Nagoya University, Furo-cho, Nagoya, Japan

Clara M. Velte Technical University of Denmark, Kgs. Lyngby, Denmark

Roel Verstappen University of Groningen, Groningen, The Netherlands

Ricardo Vinuesa Linné FLOW Centre, KTH Mechanics, Stockholm, Sweden; Swedish e-Science Research Centre (SeRC), Stockholm, Sweden

Dimitar G. Vlaykov Max Planck Institute for Dynamics and Self-Organization, Göttingen, Germany

Tom T.B. Wester ForWind, Institute of Physics, University of Oldenburg, Oldenburg, Germany

Michael Wilczek Max Planck Institute for Dynamics and Self-Organization, Göttingen, Germany

David H. Wood Department of Mechanical and Manufacturing Engineering, University of Calgary, Calgary, AB, Canada

Nobumitsu Yokoi Institute of Industrial Science, University of Tokyo, Meguro, Tokyo, Japan

Akira Yoshizawa Emeritus at Institute of Industrial Science, University of Tokyo, Tokyo, Japan

El-Sayed Zanoun Faculty of Engineering, Benha University, Benha, Egypt; Department of Aerodynamics and Fluid Mechanics, Brandenburg University of Technology, Cottbus, Germany

Part I
Theory

Emergence of Non-Gaussianity in Turbulence

Michael Wilczek, Dimitar G. Vlaykov and Cristian C. Lalescu

Abstract Fully developed turbulence is characterized by markedly non-Gaussian statistics. Here, we discuss some aspects of the relation between non-Gaussianity, the emergence of coherent structures and phase correlations in turbulence. Direct numerical simulations of homogeneous isotropic turbulence are used to demonstrate a fairly rapid emergence of non-Gaussian statistics from Gaussian initial conditions.

1 Introduction

One hallmark of fully developed turbulent flows is the intrinsic non-Gaussianity of the velocity field. For example, the single-point probability density function (PDF) of the velocity, which characterizes the large scales, remains close to Gaussian with slightly sub-Gaussian tails [1, 2]. The PDFs of small-scale quantities such as the vorticity, however, exhibit broad tails. This translates to the frequent occurrence of extreme events and is a signature of small-scale coherent structures such as vortex tubes or strain sheets. In comparison, Gaussian fields, whose multi-point statistics are jointly Gaussian, appear largely structureless.

Velocity increment PDFs effectively interpolate between the small- and large-scale statistics. Using them to probe turbulent velocity fields on increasing scales reveals a breaking of statistical self-similarity: the PDFs change shape as a function of scale. This well-known phenomenon of intermittency is absent in Gaussian fields, which are statistically self-similar. This motivates the question of how non-Gaussianity and intermittency arise from the turbulent dynamics.

In the following we give a qualitative discussion on the relation between non-Gaussian statistics and the emergence of coherent structures and phase correlations in turbulence. In particular, we study the evolution of a turbulent flow from Gaussian

M. Wilczek (✉) · D.G. Vlaykov · C.C. Lalescu
Max Planck Institute for Dynamics and Self-Organization,
Am Faßberg 17, 37077 Göttingen, Germany
e-mail: michael.wilczek@ds.mpg.de

R. Örlü et al. (eds.), *Progress in Turbulence VII*, Springer Proceedings
in Physics 196, DOI 10.1007/978-3-319-57934-4_1

initial conditions by means of direct numerical simulations. Finally, we outline how these insights may be useful to better understand and model intermittency in turbulence.

2 Direct Numerical Simulations and Gaussian Initial Conditions

We present direct numerical simulations (DNS) of statistically stationary homogeneous isotropic turbulence at a Taylor-based Reynolds number of $R_\lambda \approx 129$. A standard pseudo-spectral solver is used to simulate the Navier–Stokes equation in the vorticity formulation in a periodic box with 512^3 grid points at a resolution of $k_M \eta_K \approx 1.67$, where k_M denotes the highest resolved mode and η_K the Kolmogorov length. The flow is forced on the large scales by a term linear in the band-passed vorticity (Lundgren forcing) [3, 4]. Time-stepping is performed with a third-order memory-saving Runge–Kutta method [5].

To generate Gaussian initial conditions, a snapshot from the simulation is taken from the statistically stationary regime. The Fourier coefficients $\mathbf{u}(\mathbf{k})$ are rotated in the complex plane, $\mathbf{u}(\mathbf{k}) \rightarrow \mathbf{u}(\mathbf{k})\, \mathrm{e}^{\mathrm{i}\varphi(\mathbf{k})}$, with statistically independent random phases $\varphi(\mathbf{k})$, which are uniformly distributed in $[0, 2\pi]$ for each $\mathbf{k}$. Reality of the velocity field is imposed by ensuring $\varphi(\mathbf{k}) = -\varphi(-\mathbf{k})$. As will be clarified in Sect. 4, the resulting velocity field is close to Gaussian. Seventeen distinct realizations of $\varphi(\mathbf{k})$ are used, such that an ensemble of approximately Gaussian fields is obtained. These fields are then taken as initial conditions for new DNSs, the results of which are discussed in the following sections.

3 Emergence of Non-Gaussian Statistics

A qualitative impression of the DNS results with approximately Gaussian initial conditions can be gained from Fig. 1, which shows visualizations of the velocity and vorticity fields for three subsequent instances in time. The fact that Gaussian initial conditions appear structureless is particularly evident in the vorticity visualizations. As the flow evolves under the Navier–Stokes dynamics, small-scale coherent structures start to emerge rapidly. Already after ten Kolmogorov time scales (characterizing the fastest turbulent dynamics) they can be clearly identified. The fine-scale structure of the velocity field changes on a comparable time scale. However, its large-scale structure changes only slowly. The difference between the two fields is understood through a simple eddy turnover argument by considering that the velocity is a large-scale and the vorticity a small-scale quantity.

Quantitatively, consider the single-point PDFs for the velocity and vorticity fields (Fig. 2). The velocity PDF is initially close to Gaussian with slightly sub-Gaussian tails. The deviations can be explained by the fact that the initial velocity field is steeper than k^{-1}, which leads to corrections to the behavior expected from a central limit

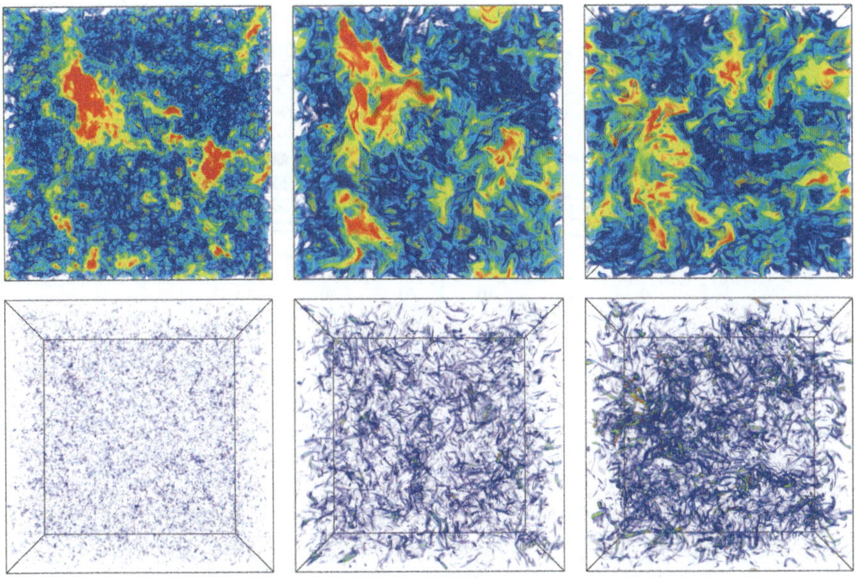

Fig. 1 Snapshots of DNS velocity fields (*top row*) and vorticity fields (*bottom row*) at simulation times of approximately 0, 10, 100 Kolmogorov time scales. Already after ten Kolmogorov time scales small-scale vorticity structures have emerged, and they grow stronger over time. Also the fine-scale structure of the velocity field changes, whereas large-scale features remain qualitatively similar

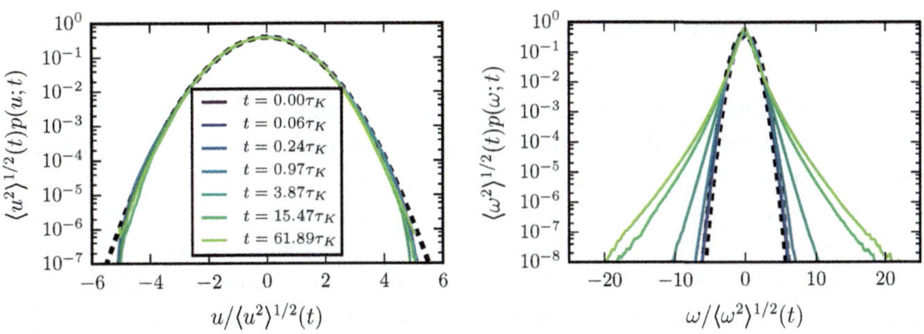

Fig. 2 One-point PDFs for velocity and vorticity fields. The velocity PDFs are slightly sub-Gaussian throughout the evolution. The vorticity PDFs start out as Gaussian, but quickly deviate and settle into a strongly non-Gaussian form

theorem argument [6]. As the field evolves from these initial conditions, the velocity PDF varies slightly due to the temporal evolution of fluctuations, but remains very close to Gaussian as expected from previous theoretical considerations [7]. Along with the emergence of small-scale coherent structures, the vorticity PDF rapidly develops heavy tails [8]. This nicely visualizes the common picture of turbulence, in which small-scale coherent structures break statistical self-similarity and therefore contribute to non-Gaussianity and intermittency.

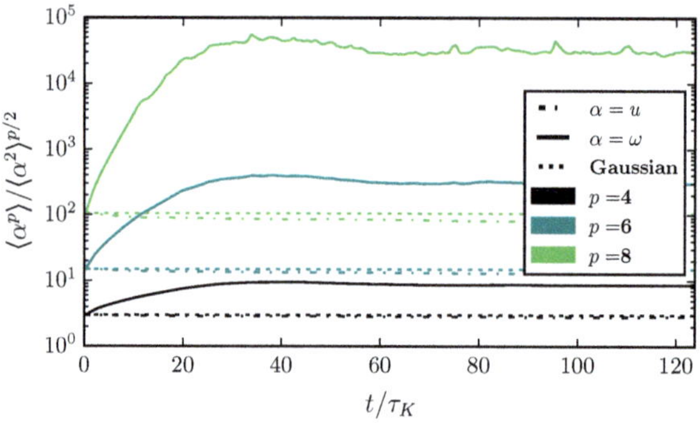

Fig. 3 Time evolution (in units of the Kolmogorov time τ_K) of the normalized moments of the velocity field (*dash-dot*) and the vorticity field (*solid*). While for the velocity field the values remain slightly sub-Gaussian, the vorticity quickly exhibits strong non-Gaussianity. Gaussian values are shown as *dashed lines* for reference

The emergence of non-Gaussian statistics from Gaussian initial conditions can also be confirmed by studying moments of the velocity and vorticity fields as a function of time, as presented in Fig. 3. As can be seen, the single-point moments of the velocity field stay close to the Gaussian values. Consistent with the observations of the vorticity PDF, the single-point vorticity moments rapidly depart from their Gaussian initial values.

4 Phase Correlations

There is an intimate relation between Gaussianity and phase correlations. For a Gaussian random field different Fourier modes are mutually statistically independent. One should note that this is a much stronger statement (as it pertains to all statistical moments) than the fact that for homogeneous random fields the phases are uncorrelated (which concerns only second-order moments). As mentioned above, the opposite also holds: A superposition of Fourier amplitudes with random phases results in approximately Gaussian statistics under quite general conditions (see e.g. [9]). Thus phase correlations are a signature of non-Gaussian statistics.

To explain the emergence of phase correlations, it is instructive to consider the well-known Fourier representation of the Navier–Stokes equations for the amplitude $a_l(\mathbf{k}, t)$ and the phase $\varphi_l(\mathbf{k}, t)$:

$$\begin{aligned}
\left(\partial_t + \nu k^2\right) a_l(\mathbf{k}) &= \frac{1}{2} \sum_{m,n} P_{lmn}(\mathbf{k}) \sum_{\mathbf{p}+\mathbf{q}=\mathbf{k}} a_m(\mathbf{p}) a_n(\mathbf{q}) \sin\left[\varphi_m(\mathbf{p}) + \varphi_n(\mathbf{q}) - \varphi_l(\mathbf{k})\right] \\
-a_l(\mathbf{k})\, \partial_t \varphi_l(\mathbf{k}) &= \frac{1}{2} \sum_{m,n} P_{lmn}(\mathbf{k}) \sum_{\mathbf{p}+\mathbf{q}=\mathbf{k}} a_m(\mathbf{p}) a_n(\mathbf{q}) \cos\left[\varphi_m(\mathbf{p}) + \varphi_n(\mathbf{q}) - \varphi_l(\mathbf{k})\right] .
\end{aligned} \tag{1}$$

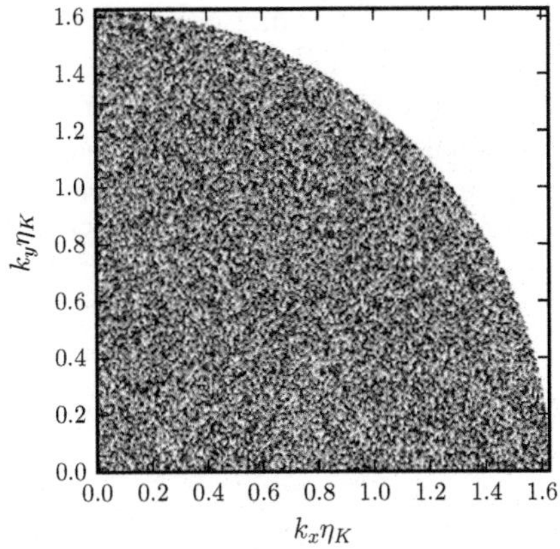

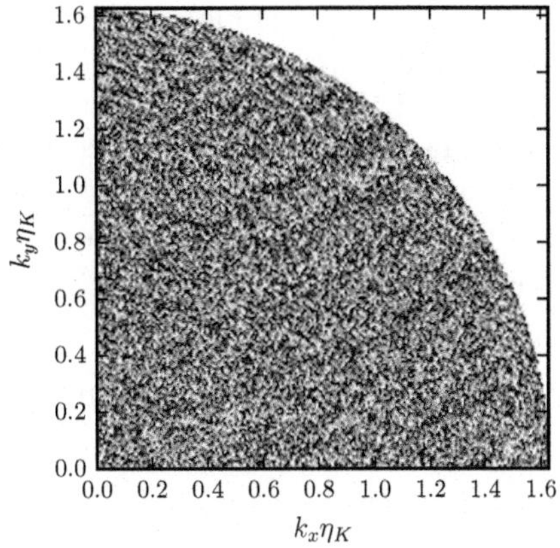

Fig. 4 Phases of one component of the velocity field in the $k_z\eta_K = 0$ plane. The *left plot* shows the phases for one of the Gaussian initial conditions, while the *right plot* shows the phases for the same velocity field after the quasi-stationary regime has been reached

Here, $P_{lmn}(\mathbf{k}) = k_m P_{ln}(\mathbf{k}) + k_n P_{lm}(\mathbf{k})$ is a suitably defined projection operator based on $P_{lm}(\mathbf{k}) = \delta_{lm} - \frac{k_l k_m}{k^2}$ [10]. It is interesting to note that the linear (viscous) term does not impact the phase dynamics directly. In contrast, the nonlinear term couples phase triads of Fourier coefficients which fulfill $\mathbf{k} = \mathbf{p} + \mathbf{q}$. This coupled phase dynamics then gives rise to correlations among phases.

Figure 4 shows the phases of a single velocity component in the $k_z\eta_K = 0$ plane. The left panel corresponds to the close to Gaussian initial conditions. By construction, all of the phases are statistically independent. The right panel shows the phases after approximately $76\tau_K$. As expected for turbulence, the phases are more or less random, however there is a remarkable degree of visual coherence, especially at smaller scales, with approximately equidistant striations visible in the "1 o'clock" region of the quadrant.

5 Conclusion and Outlook

By investigating numerical simulations of fully developed turbulence from approximately Gaussian initial conditions, we have seen that small-scale statistics (like the vorticity PDF) rapidly transition to non-Gaussianity whereas the large scales (exemplified by the single-point velocity PDF) remain close to Gaussian. In Fourier space, this is accompanied by the emergence of phase correlations, which appear more pronounced at smaller scales. Combining these two observations, our current working hypothesis is that intermittency in turbulence can be interpreted in terms of scale-dependent phase correlations. Due to the complexity of the Navier–Stokes equations, it remains a formidable task for future work to derive first-principle results on such phase correlations.

Meanwhile, it may turn out to be useful to study toy models. For example, one may consider the temporal evolution of a one-dimensional periodic field with the Fourier representation $u(x,t) = \sum_k a(k)e^{\mathrm{i}[\varphi(k,t)+kx]}$. Assuming that the amplitudes remain fixed in time, one can impose a phase dynamics reminiscent of the Navier–Stokes dynamics (1):

$$\dot{\varphi}(k) = \sum_{p=-N}^{N} \omega(k,p) \cos\left[\varphi(p) + \varphi(k-p) - \varphi(k)\right] \tag{2}$$

where $\omega(k,p) = -\frac{ka(p)a(k-p)}{a(k)}$ are coupling coefficients depending on the prescribed Fourier amplitudes. Note that this phase coupling model is even simpler than the one-dimensional Burgers equation, as the amplitudes here are time-independent. We refer to [11] for an analysis of the phase dynamics in the one-dimensional Burgers equation. First numerical results indeed confirm that the phase coupling model (2) displays phase correlations which are more pronounced at smaller scales. As a consequence, small-scale statistics of the resulting field in real space depart more strongly from Gaussianity than large-scale statistics. Such toy models could also be useful in the modeling of intermittency, their simplicity inviting more direct analytical approaches. It is interesting to note that the phase coupling model establishes a relation to the field of non-locally coupled oscillators which are known to display a plethora of dynamical states including phase synchronization, chaos and chimera states. It will be exciting to see to which extent such concepts also apply to the phase dynamics of turbulence.

Acknowledgements We acknowledge insightful discussions with Michele Buzzicotti and Miguel Bustamante. This work was supported by the Max Planck Society.

References

1. J. Jiménez, A.A. Wray, P.G. Saffman, R.S. Rogallo, The structure of intense vorticity in isotropic turbulence. J. Fluid Mech. **255**, 65–90 (1993)
2. A. Noullez, G. Wallace, W. Lempert, R.B. Miles, U. Frisch, Transverse velocity increments in turbulent flow using the RELIEF technique. J. Fluid Mech. **339**, 287–307 (1997)
3. T.S. Lundgren, Linearly forced isotropic turbulence in *Annual Research Briefs* (Center for Turbulence Research, Stanford), pp. 461–473 (2003)
4. C. Rosales, C. Meneveau, Linear forcing in numerical simulations of isotropic turbulence: physical space implementations and convergence properties. Phys. Fluids **17**(9), 095106 (2005)
5. C.-W. Shu, S. Osher, Efficient implementation of essentially non-oscillatory shock-capturing schemes. J. Comput. Phys. **77**(2), 439–471 (1988)
6. J. Jiménez, Turbulent velocity fluctuations need not be Gaussian. J. Fluid Mech. **376**, 139–147 (1998)
7. M. Wilczek, A. Daitche, R. Friedrich, On the velocity distribution in homogeneous isotropic turbulence: correlations and deviations from Gaussianity. J. Fluid Mech. **676**, 191–217 (2011)

8. M. Wilczek, R. Friedrich, Dynamical origins for non-Gaussian vorticity distributions in turbulent flows. Phys. Rev. E **80**, 016316 (2009)
9. J.M. Bardeen, J.R. Bond, N. Kaiser, A.S. Szalay, The statistics of peaks of Gaussian random fields. Astrophys. J. **304**, 15–61 (1986)
10. R.H. Kraichnan, The structure of isotropic turbulence at very high Reynolds numbers. J. Fluid Mech. **5**(4), 497–543 (1959)
11. M. Buzzicotti, B.P. Murray, L. Biferale, M.D. Bustamante, Phase and precession evolution in the Burgers equation. Eur. Phys. J. E **39**(3), 34 (2016)

Percolation: Statistical Description of a Spatial and Temporal Highly Resolved Boundary Layer Transition

Tom T. B. Wester, Dominik Traphan, Gerd Gülker and Joachim Peinke

Abstract In this article spatio-temporally resolved particle image velocimetry data of a flat plate's boundary layer are shown. With this set up, it is possible to capture the highly unsteady phase transition from laminar to turbulent state of the boundary layer close to the surface. In the evaluation of the boundary layer data it is shown that it is possible to link the laminar-turbulent phase transition to the (2+1)D directed percolation universality class. This can be shown by the unique exponents of the directed percolation class which will be extracted from the PIV data.

1 Introduction

The description of transition into turbulence has always been a challenging task. Thirty years ago Pomeau was the first to describe the dynamics of laminar-turbulent transition by a system of coupled oscillators [1]. Thereby he paved the way for the statistical description of laminar-turbulent transition by the directed percolation theory. This theory allows a simple description of complex phase transitions with only three critical exponents. These exponents are unique for each universality class of percolation, so the transition from a laminar to a turbulent flat plate's boundary layer may be ascribed to a known class.

Until the last decade it was not possible to provide experimental evidence to show the spatio-temporal intermittency which occurs in the transition from laminar to turbulent flow. Due to more accurate measurement techniques nowadays it is possible to capture the transition with much higher temporal and spatial resolution.

This has led to more detailed investigations with respect to directed percolation of different flow situations such as channel flow [2, 3], Couette flow [4], shear flows [5–8] and fully turbulent flows [9]. All of them show promising results, which support the presumption of Pomeau.

T.T.B. Wester (✉) · D. Traphan · G. Gülker · J. Peinke
ForWind, Institute of Physics, University of Oldenburg, Oldenburg, Germany
e-mail: tom.wester@uol.de

R. Örlü et al. (eds.), *Progress in Turbulence VII*, Springer Proceedings
in Physics 196, DOI 10.1007/978-3-319-57934-4_2

In contrast to the experiments mentioned, the evaluation presented in this article is the first which is carried out on a flat plate's boundary layer in an airflow and also the first evaluation which uses PIV data as basis.

2 Experimental Setup and Methodology

The experiment is based on high speed stereoscopic particle image velocimetry (HS-PIV). In order to investigate the boundary layer of the flat plate, a wind tunnel with a closed test section is used (see Fig. 1). It has a cross section of 25×25 cm^2 and a length of 200 cm. The flat plate under investigation has a length of 100 cm. It is positioned 30 cm downstream the wind tunnel nozzle at a height of 10 cm above the bottom of the test section.

The experiment is performed at a velocity of $u_\infty = 11.5$ m s^{-1} and a free stream turbulence intensity below 0.3%. Due to the limited field of view (FOV) of the PIV measurements it is necessary to induce perturbations into the boundary layer. Therefore a small step of 0.1 mm height is placed 5 cm downstream the leading edge. Hereby the transition area can be tuned so that its streamwise length fits into the HS-PIV FOV.

The used HS-PIV system consisted of a high speed laser LDY 303 by Litron, light sheet optics and two Phantom Miro M320S high-speed cameras which are used at a reduced resolution of 1408×1048 px^2 resulting in 154×126 PIV interrogation windows. The light sheet is directed perpendicular to the inflow direction and illuminates a plane parallel to the plate's surface. The origin of the plane is located 5.8 cm downstream of the leading edge. In focus the light sheet has a thickness of 1 mm and

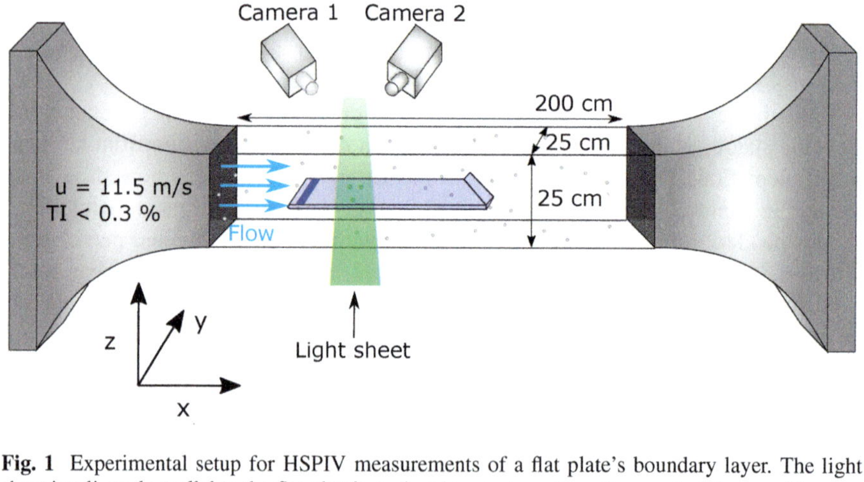

Fig. 1 Experimental setup for HSPIV measurements of a flat plate's boundary layer. The light sheet is adjusted parallel to the flat plate's surface in a region where the onset of the transition can be captured

its height above the plate's surface is 1 mm. The boundary layer thickness estimated by the Blasius equation is 2 mm at the point of measurement. Thus, the upper half of the boundary layer is investigated in the PIV measurement.

With this set up a sampling rate of 1000 velocity fields per second is possible. The velocity fields have a size of 6 cm in streamwise (x-direction) and 5 cm in spanwise direction (y-direction) with a spatial resolution of 0.04 cm in each direction. In total, 8250 velocity fields in three independent measurements are acquired. Each measurement has a duration of 2.75 s, because of the limited internal camera storage. Particle images are captured and processed using LaVision software DaVis 8.3.

3 Experimental Data

In the percolation theory only two states exist: a cell is either laminar (off) or turbulent (on). For this reason, all velocity fields need to be binarized by a certain criterion. In Fig. 2a snapshot of the velocity magnitude is plotted. Shown is the development of the velocity along the local Reynolds number $Re_x = \frac{u(x)\cdot x}{\nu}$ where x is the denoted as the propagation length along the plate and the spanwise direction. The right hand side shows the same velocity field, but binarized.

In order to binarize the data, a velocity threshold is used. Based on the velocity, one can directly draw a conclusion on the state of the boundary layer. If the velocity increases compared to the laminar boundary layer there must be a mixing between the high energetic ambient flow and the low energetic boundary layer. This only happens if the boundary layer becomes unstable and thus transitive and turbulent.

According to that, the interrogation cell is set to 0 (off) if $u(x)_{TH}$ and 1 (on) otherwise. A parameter variation yielded $u_{TH} = 4\ \text{m s}^{-1}$ as an appropiate threshold that reflects turbulent structures for the given measurement distance to the wall. However, percolation exponents do only slightly depend on chosen velocity threshold.

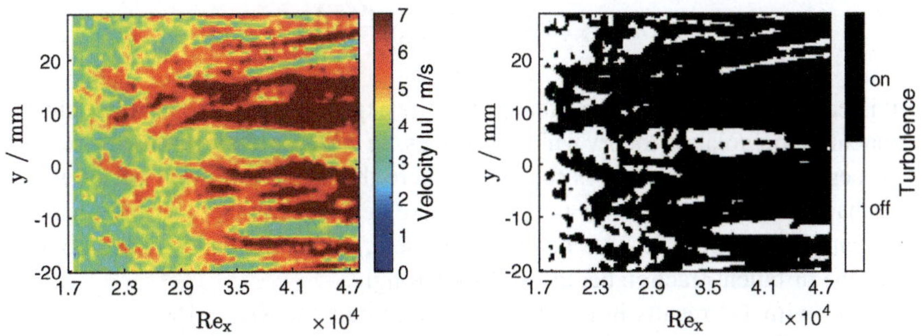

Fig. 2 *Left* Example of a measured velocity field. *Right* Binarized velocity field by a threshold of $u_{Thresh} = 4\ \text{ms}^{-1}$

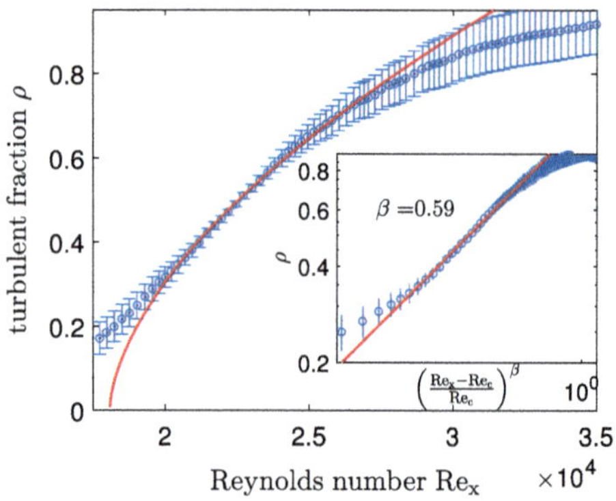

Fig. 3 Scaling of turbulent fraction with Reynolds number $Re_x = \frac{u(x) \cdot x}{\nu}$ in streamwise direction. The measurement points are shown in *blue* with *error bars*. The *red curve* represents a fit according to (1). The *inset* shows the plot in logarithmic scale to emphasize the exponential behavior. Here, the turbulent fraction is plotted against the reduced Reynolds number

The turbulent fraction $\rho(Re_x)$ can be determined directly from the binarized data. The turbulent fraction is the ratio of turbulent to laminar cells for each interrogation window. For this reason, a two-dimensional turbulent fraction results for the PIV data. This two-dimensional turbulent fraction is averaged along the y-direction to obtain a one-dimensional distribution. This distribution is plotted against the local Reynolds number Re_x in Fig. 3. The shown errorbars represent the standard deviation. The inset of the figure shows the logarithmic illustration of the turbulent fraction over the reduced Reynolds number $\frac{Re_x - Re_c}{Re_c}$. The turbulent fraction shows a monotonic growth with increasing local Reynolds number. At the so called critical point the turbulent fraction increases rapidly. This is typical for directed percolation and can be described by the exponential relation between the turbulent fraction $\rho(Re_x)$, the local Reynolds number Re_x and the critical point Re_c shown in (1):

$$\rho(Re_x) = \rho_0 \cdot \left(\frac{Re_x - Re_c}{Re_c} \right)^{\beta}. \tag{1}$$

In this equation ρ_0 is a constant factor and the exponent β is one of three unique exponents of percolation theory which describes the increase of the turbulent fraction.

In percolation theory, Re_c is the point where the phase transition between two states happens. In our case, it is the point where the transition between laminar and turbulent phase begins. In case of the experiment, this corresponds to the point at which the turbulent fraction begins to grow strongly.

The fit with (1) results in a critical Reynolds number $Re_c = 18040 \pm 380$ and the exponent $\beta_{exp} = 0.59 \pm 0.04$. In order to obtain the other two unique exponents of percolation theory $\mu_{\parallel}$ and $\mu_{\perp}$ which describe the spatial and temporal spreading behavior, the development of laminar clusters must be considered. Laminar cluster are regions in the time development of the flow at Re_c where laminar cells are

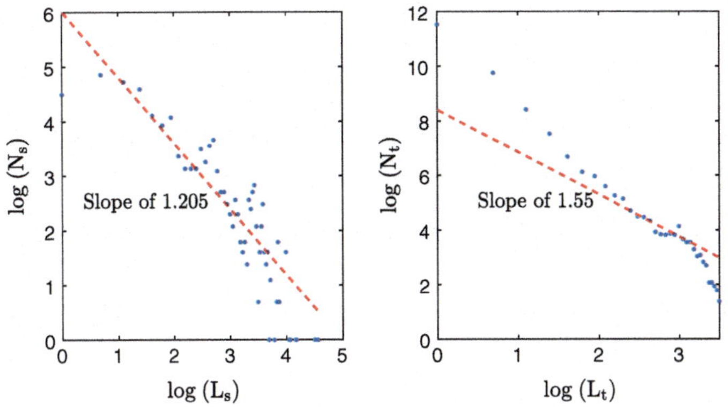

Fig. 4 *Left* Histogram of laminar clusters in space. *Red dashed line* represents the theoretical slope of (2+1)D directed percolation. *Right* Histogram of laminar clusters in time. *Red dashed line* represents the theoretical slope of (2+1)D directed percolation

surrounded by turbulent cells. In percolation theory, the number ($N_{s,t}$) of clusters of a particular size ($L_{s,t}$) scale with the following laws for space (2) and time (3):

$$N_s(L_s) \propto L_s^{\mu_\perp} \tag{2}$$

$$N_t(L_t) \propto L_t^{\mu_\parallel}. \tag{3}$$

The histograms of the cluster sizes are shown in Fig. 4. On the left hand side, the laminar cluster sizes in space, and on the right hand side, the cluster sizes in time are shown. The red dashed lines show slopes of the theoretical cluster distribution in (2+1)D directed percolation theory.

4 Discussion and Concluding Remarks

In the preceding chapter, the experimental results were presented. From the turbulent fraction, the first unique exponent β is derived. This exponent is close to the theoretical exponent for (2+1)D directed percolation $\beta_{theo} = 0.583$. The other unique exponents for the scaling behavior of the cluster sizes also follow the laws of (2+1)D directed percolation visually. The cluster in space follow the law better than the clusters in time. The small deviation could be a result of the limited temporal resolution. Another explanation is a limited size effect due to the limited FOV of the PIV measurements.

All in all, the transition of the flat plate's boundary layer can be described by the (2+1)D directed percolation theory. The extracted exponents from the experiment seem to match the theory. Therefore, the highly dynamic behavior of the transition can be reduced to mainly three unique exponents and this phenomenon can be assigned to one of the known percolation classes.

5 Future Work

Since these first results are promising a validation is needed. Therefore, the turbulent boundary layer of the flat plate will be investigated more detailed at different heights and at different free stream velocities.

Furthermore, other transient flows such as channel flow or even more complex flows such as airfoil boundary layers will be investigated. If it is possible to show that transitions from laminar to turbulent flow belong to a universality class of percolation this results could be used for computational fluid dynamic models to simulate the transition in a more accurate way.

Acknowledgements The authors thank the Niedersächsische Ministerium für Wissenschaft und Kultur (Ministry of Science and Culture of Lower Saxony) for financial support.

References

1. Y. Pomeau, Front motion, metastability and subcritical bifurcations in hydrodynamics. Physica D: Nonlinear Phenomena **23**(1), 3–11 (1986)
2. D. Barkley, Simplifying the complexity of pipe flow. Physical Review E **84**(1), 016309 (2011)
3. M. Sano, K. Tamai. A universal transition to turbulence in channel flow. *Nature Physics*, 2016
4. G. Lemoult, L. Shi, K. Avila, S.V Jalikop, M. Avila, B. Hof. Directed percolation phase transition to sustained turbulence in couette flow. *Nature Physics*, 2016
5. T. Korinna, Allhoff and Bruno Eckhardt. Directed percolation model for turbulence transition in shear flows. Fluid Dynamics Research **44**(3), 031201 (2012)
6. T. Kreilos, B. Eckhardt, T.M Schneider. Increasing lifetimes and the growing saddles of shear flow turbulence. *Physical review letters*, **112**(4):044503, 2014
7. H.-Y. Shih, T.-L. Hsieh, N. Goldenfeld. Ecological collapse and the emergence of travelling waves at the onset of shear turbulence. *Nature Physics*, 2015
8. M. Avila, B. Hof, Nature of laminar-turbulence intermittency in shear flows. Physical Review E **87**(6), 063012 (2013)
9. D. Barkley, B. Song, V. Mukund, G. Lemoult, M. Avila, B. Hof, The rise of fully turbulent flow. Nature **526**(7574), 550–553 (2015)

The Key Role of Pressure in the Turbulence Cascading Process

Patrick Bechlars and Richard D. Sandberg

Abstract This work revisits the cascading mechanism in turbulence from a novel perspective on the scale transfer and highlights the key role of pressure within this concept. The turbulent cascade is the core of all complexity in a turbulent flow. A better understanding of this process is essential for complexity reducing turbulence models that are essential for predictive simulations in engineering and the natural sciences. Here, the significance of the pressure within the turbulence dynamics is highlighted and it is shown how information about the pressure can be used to interpret and explain certain features of a turbulent flow. It has been shown previously how the presence of a wall affects the dynamics of turbulence. The present work explains how these dynamics are linked to the cascading mechanism that drives turbulence. The conclusions are believed to be transferable to other turbulent flows.

1 Introduction

Turbulence has been the subject of research for more than a century. In that time significant advances had been made in the understanding of turbulent flows. Nevertheless, many details are still not understood well enough. Some of those aspects are, e.g., the role of pressure and the multiple mechanisms that contribute to the energy cascade transferring energy between the different scales of motion.

Gotoh and Nakano [1] investigated the role of pressure in turbulent flows by conditioning homogeneous isotropic turbulence data obtained through direct numerical simulation (DNS) with pressure. They summarized that the pressure is a key for scaling the velocity structure functions and is important for some of their dynamics. Wilczek and Meneveau [2] give a novel insight into the dynamics of the velocity gradient tensor and discuss the role of pressure in these dynamics. However, the

P. Bechlars (✉) · R.D. Sandberg
University of Melbourne, Parkville, Australia
e-mail: patrick.bechlars@gmail.com
URL: http://www.bechlars.org

R.D. Sandberg
e-mail: richard.sandberg@unimelb.edu.au

R. Örlü et al. (eds.), *Progress in Turbulence VII*, Springer Proceedings in Physics 196, DOI 10.1007/978-3-319-57934-4_3

link to actual physical structures remains open in their work. This gap was recently closed by Lawson and Dawson [3] who provided new insight into the arrangement of turbulent structures and discussed some aspects of the role of pressure. Such insights are particularly helpful for understanding the non-local effects within the turbulence dynamics. Bechlars and Sandberg [4] have shown how the presence of a wall affects the dynamics of turbulence and have highlighted that the understanding of the role of pressure in the turbulence dynamics is crucial for a better understanding of turbulence in general. In another work they have highlighted that there is a peak in the frequency of occurrence of unstable vortical structures in the near-wall region of a turbulent boundary layer (TBL) [5]. Further, their results suggest that a reverse cascade might be present with the same flow regime of unstable vortical structures in the entire TBL.

In this work we discuss the role of pressure as a stabiliser of vortical structures and how it looses that ability in the near-wall region. By applying the critical point concept introduced by Perry and Chong [6] as a condition on the transfer of energy between different scales of motion, the suggestions of [5] about the reverse cascade are supported. The physical mechanism for this process is hypothesised with similar arguments about the pressure that are applied when describing the pressure as a stabiliser. The results are based on the DNS data of a TBL [5].

The necessary background for the discussion about the pressure is given in Sect. 2. The results are presented and discussed in Sect. 3 and conclusions are presented in Sect. 4.

2 Method

For a general understanding about the role of pressure p it is helpful to consider the inviscid form of the momentum transport equation in Lagrangian form, i.e. we follow a fluid particle,

$$\rho \frac{d\mathbf{u}}{dt} = -\nabla p. \tag{1}$$

Here ρ is the density, $\mathbf{u}$ is the velocity vector and $\frac{d}{dt}$ denotes the material derivative. In this form the similarity of the momentum transport to the equation of motion for a particle becomes obvious. For now we interpret a fluid particle as a particle in motion that is governed by the single external force which is described by the pressure gradient $-\nabla p$. If we take the divergence of the incompressible momentum equation, this leads to the equation that describes the Laplacian of pressure as a function of the second velocity gradient tensor invariant $Q = -\frac{1}{2}\text{trace}(AA)$ (with A being the velocity gradient tensor)

$$-\Delta p = -2Q. \tag{2}$$

This is a Poisson equation and with a simplification, that the considered domain is unbounded in $\mathbb{R}^3$ the general solution for this pressure Poisson equation is

$$p(\mathbf{x}) = \iiint_{\mathbb{R}^3} -2Q(\mathbf{y}) \frac{1}{4\pi} \frac{1}{\|\mathbf{x}-\mathbf{y}\|} d\mathbf{y}. \quad (3)$$

The assumption of the domain being unbounded is reasonable if turbulent structures are discussed that are relatively small compared to their distance to the wall. In case of getting closer to a wall a second convolution that incorporates the pressure boundary condition needs to be added to the solution. The purpose of the current work is to introduce the idea that pressure can be seen as a convolution of the fundamental solution of the Poisson equation with $-2Q$. This means that the pressure can be seen as a filtered version of the Q-field with the fundamental solution as filter kernel. Hence, this assumption is acceptable for the initial observations in this work.

3 Results and Discussion

In the outer region of the TBL the pressure field aligns well with the vortical turbulent structures as visually supported through Fig. 1, left. This arrangement is changed in the near-wall region (Fig. 1, center, right) where the pressure iso-surfaces form a hull around the turbulent structures, represented by the Q-criterion. This behaviour can be explained by the pressure being a low-pass-filtered representation of the Q-field as presented in (3). The pressure field represents dense clusters of single turbulent structures, found in the near-wall region, as an enveloping larger structure. Further, the pressure interacts visibly different with the wall than the velocity field. The pressure structures create a clear foot at the wall, whereas the Q-structures do not reach the wall.

This variation of alignment of the pressure and the Q-field across the TBL is well quantified by the p.d.f. of the alignment between the gradient of the second

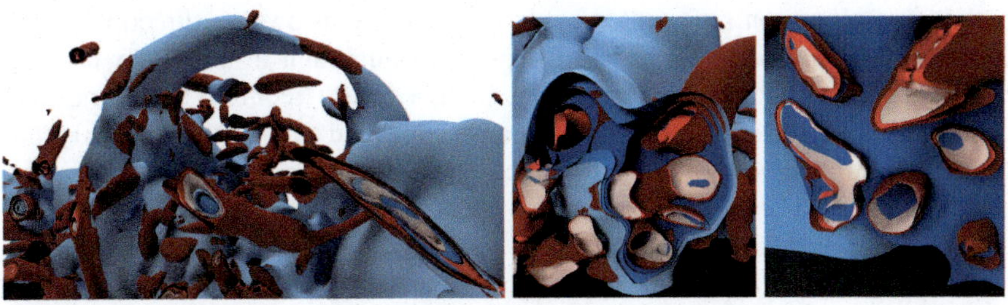

Fig. 1 Turbulent structures above the wall (*dark grey*) in a TBL are shown via iso-surfaces of different levels of Q (*shades of red*) and pressure (*shades of blue*). The *left* image ($y^+ \approx 115 - 370$) shows regions further away from the wall whereas the two images on the right ($y^+ \approx 0 - 80$) show regions closer to the wall

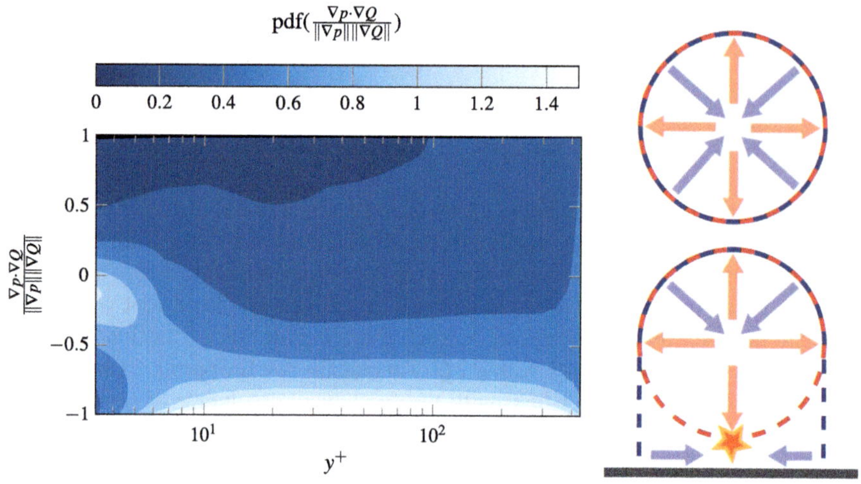

Fig. 2 *Left* P.d.f.s of the alignment of ∇p and ∇Q for different wall-normal locations. *Right* Cross sections of a schematic vortex shown via an iso-line of Q (*red dashed*) and p (*blue dashed*). The centrifugal force acting on a fluid particle following the iso-lines is shown as *red arrows* and the force applied by the pressure is shown as *blue arrows*. *Right/top* The vortex is free of any boundary conditions. *Right/bottom* The vortex' pressure field is affected by the wall. The most unstable location is marked with a spark

velocity gradient tensor invariant Q and the pressure gradient (Fig. 2, left). Whereas both vectors prefer a negative parallel alignment above $y^+ = 20$ (neglecting some discrepancies due to intermittency in the outer layer), they transition towards a preferred perpendicular alignment when approaching the wall. We know from literature (e.g. [7]), that vortices have high values of Q in their core whereas the fluid is swirling around this local maximum of Q. From the filtered character of pressure (3) one can see that the vortex core carries a local pressure minimum as well, as previously reported in the literature (e.g. [8]). In Fig. (2, right) the cross section of a schematic vortex is shown via iso-lines of pressure as well as iso-lines of Q. The red arrows (in direction of $-\nabla Q$) represent the centrifugal forces that are caused by the vortical motion represented by Q and the blue arrows (in direction $-\nabla p$) represent the pressure forces due to the pressure minimum in the vortex core. In the top image, one can see the forces that act in a vortex free from boundaries and interactions. Going back to the particle idea we know that a moving particle does not stay on a circular movement without forces acting on it. Equation (1) states that the pressure force (blue) will act against the centrifugal force (red) and therefore keeps the particle on its circular trajectory. This schematic view of vortices illustrates that the pressure forces stabilize rotational motions and allow vortices to exist. On the other hand, if we move towards a wall or in clusters of structures then we have shown that this stabilising alignment of the pressure with the vortex is disturbed. This leads to an enhanced breakup of vortices as illustrated in the bottom schematic, and was previously [5] detected in the present TBL data.

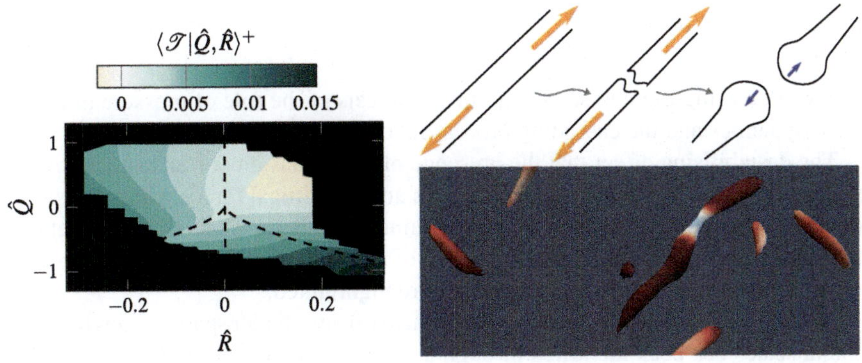

Fig. 3 *Left* Conditional averages of the energy transfer between large spatial scales and small spatial scales based on a filter width of $\Delta^+ = 30$ at $y^+ \approx 339$, conditioned with the respective QR-states. *Right/top* Schematic description of the backscatter mechanism with Vortex tube (*black*). The *arrows* indicate the stretching direction (*yellow*) and the direction of the pressure force (*blue*). *Right/bottom* Q-iso-surfaces above a wall (*grey*) colored with energy transfer: LS to SS (*red*) versus SS to LS (*blue*)

Another feature of turbulence that is influenced by pressure forces is a backscatter mechanism that transfers energy, opposite to the direction of the net energy cascade, from smaller spatial scales to larger spatial scales. Such a mechanism was identified in the present TBL data by splitting the spatial scales into large scales (LS) and small scales (SS) using a Gaussian filter with a filter width of $\Delta^+ = 30$, scaled with wall-units. By conditioning the resulting energy transfer $\mathcal{T}$ between the separated scales with the second and third velocity gradient tensor invariant Q and R (Fig. 3, left) a mean energy transfer from SS to LS can be found for a part of the structures in the regime of unstable vortical structures. This supports the suggestion made by [5] that a potential backscatter mechanism can be found in this characteristic regime. A definition of the structure types in the QR-space can be found in [6].

A hypothesis on how this backscatter mechanism physically acts is illustrated in Fig. (3, right/top). The stretching mechanism that vortices are exposed to becomes unstable after acting long enough on the vortex. In unstable regions the vortex is not stretched further and the rotation rate decays (commonly described as vortex breakup). The reduced rotation results in a local imbalance of pressure force and centrifugal force at the breakup location. Therefore the still intact parts of the vortex attract fluid due to the pressure forces caused by the remaining pressure minimum in the core now being exposed to less rotational fluid that does not counteract this force. This entrains fluid into the vortex core causing the formation of bulky ends and hence larger spatial scale. Although a rigorous validation of this hypothesis is still outstanding, a visual investigation of vortical structures in a TBL supports this mechanism. In Fig. (3, right/bottom) a vortical structure is captured during the apparent breakup process. The contours indicate that while the intact parts of the structure transfer energy from LS to SS, the part going through the breakup mechanism transfers energy from SS to LS and, hence, reflects a backscatter mechanism.

4 Conclusions

This work has highlighted some significant aspects of the role of pressure in turbulence dynamics and the cascading process in a wall-bounded flow.

The destabilising effect that the presence of a wall has on vortical structures are explained by the miss-alignment of pressure and centrifugal forces within the structures. As potential reasons for this miss-alignment the dense clustering of structures in the near-wall region as well as tendency of the pressure to interact with the wall in a different way than the velocity field were highlighted.

A backscatter mechanism that transfers energy from smaller spatial scales to larger spatial scales has been identified through a characteristic decomposition within the regime of unstable vortical structures. It is hypothesised that this backscatter is caused by the formation of bulky ends in those structures undergoing a breakup process. This process is explained via the pressure balances in the vortical structures. The hypothesis is confirmed visually, however, a rigorous elaboration remains outstanding.

As a general conclusion we want to emphasize that this work has highlighted that the understanding of the behaviour and the role of pressure significantly contribute to the understanding of turbulence. Further, it is believed that the findings for the near-wall region can be be transferred to a good extend to other flow topologies with densely packed turbulence such as strong free shear layers.

References

1. T. Gotoh, T. Nakano, Role of pressure in turbulence. J. Stat. Phys. **113**(5–6), 855–874 (2003)
2. M. Wilczek, C. Meneveau, Pressure Hessian and viscous contributions to velocity gradient statistics based on Gaussian random fields. J. Fluid Mech. **756**, 191–225 (2014)
3. J.M. Lawson, J.R. Dawson, On velocity gradient dynamics and turbulent structure. J. Fluid Mech. **780**, 60–98 (2015)
4. R.D. Sandberg, B.J. Tester, Mach-number scaling of individual azimuthal modes of subsonic co-flowing jets. J. Fluid Mech. **793**, 209–228 (2016)
5. P. Bechlars. Comprehensive characterisation of turbulence dynamics with emphasis on wall-bounded flows. PhD thesis, University of Southampton (2015)
6. A.E. Perry, M.S. Chong, A description of eddying motions and flow patterns using critical-point concepts. Ann. Rev. Fluid Mech. **19**, 125–155 (1987)
7. P. Chakraborty, S. Balachandar, R.J. Adrian, On the relationships between local vortex identification schemes. J. Fluid Mech. **535**, 189–214 (2005)
8. J. Jeong, F. Hussain, On the identification of a vortex. J. Fluid Mech. **285**(1), 69 (1995)

Lagrangian Intermittency Based on an Ensemble of Gaussian Velocity Time Series

Laura J. Lukassen and Michael Wilczek

Abstract We show that Lagrangian intermittency in fully developed turbulence can be captured in terms of an ensemble of Gaussian velocity time series. This is achieved by letting the individual ensemble members vary with respect to their correlation function. We briefly discuss how this can be analytically captured in terms of a suitably defined characteristic functional. Moreover, we present a numerical implementation of the ensemble showing a continuous change from Gaussian to non-Gaussian increment distributions for a decreasing time lag. In an outlook we show first results on the application to data from direct numerical simulation.

1 Introduction

Studying Lagrangian tracer particles in homogeneous isotropic turbulence gives fundamental insight into the nature of turbulent flows as tracer particles naturally sample the spatio-temporal complexity of fully developed turbulence. Among the simplest statistical measures to quantify this complexity are probability density functions (PDFs) of velocity increments. These PDFs display intermittency, i.e. they transition from Gaussian behavior at large time lags to non-Gaussian behavior with decreasing time lags. It is a central goal of statistical turbulence theory to explain this phenomenon. To fully capture the multiscale statistics of turbulence, however, more complex statistical quantities are needed. This, for example, includes joint statistics of Lagrangian increment statistics on various time scales and joint statistics of acceleration and velocity increments.

It is well known that dimensional analysis in the spirit of Kolmogorov 1941 phenomenology [1, 2] fails to capture Lagrangian intermittency. Under the assumption that the statistics depend only on the average rate of kinetic energy dissipation in the

L.J. Lukassen (✉) · M. Wilczek
Max Planck Institute for Dynamics and Self-Organization, Am Fassberg 17, 37077 Göttingen, Germany
e-mail: laura.lukassen@ds.mpg.de

M. Wilczek
e-mail: michael.wilczek@ds.mpg.de

R. Örlü et al. (eds.), *Progress in Turbulence VII*, Springer Proceedings in Physics 196, DOI 10.1007/978-3-319-57934-4_4

inertial range, the velocity increment PDFs turn out to be self-similar in scale. This self-similarity can be broken by replacing the mean dissipation rate in the theory with a fluctuating dissipation rate, which has been proposed by Kolmogorov and Oboukhov [3, 4] in terms of the refined similarity hypothesis. Originally this theory, nowadays known as K62, was derived for Eulerian statistics, but can also be applied to Lagrangian statistics, cf. [5]. Another successful approach in capturing Lagrangian intermittency is the multifractal framework [6, 7]. In its probabilistic interpretation, anomalous scaling of structure functions is obtained by time-lag dependent superposition of scaling laws.

In this paper, we present a general approach based on the characteristic functional which contains all statistical information along a trajectory and in this sense provides any joint statistics of interest. Since Gaussian characteristic functionals can be treated analytically, we generate non-Gaussian ensemble statistics by superimposing Gaussian characteristic functionals. From a conceptual point of view, our model may be regarded as a generalization of the above-mentioned K62 and multifractal approach as well as the work by Castaing et al. [8] who superimposed Gaussian PDFs with varying variance. In Sect. 2, an introduction to our model is given. In Sect. 3, we show through a numerical evaluation of an ensemble of Gaussian trajectories that this model is capable of producing intermittency. This proof of concept is the essential part of this paper. As a next step, the model will be applied to turbulence data from direct numerical simulation (DNS), preliminary results are presented in Sect. 4.

2 An Ensemble of Gaussian Characteristic Functionals

The complete statistical information along a Lagrangian trajectory can be described in terms of the characteristic functional

$$\varphi[\alpha] = \left\langle \exp\left(\mathrm{i} \int_{-\infty}^{\infty} \mathrm{d}t \alpha(t) u(t) \right) \right\rangle, \tag{1}$$

where the angular brackets denote an ensemble average. Equation (1) represents the inverse Fourier transform of the probability density functional of $u(t)$. Here, $u(t)$ is the velocity time series along a Lagrangian trajectory, and $\alpha(t)$ denotes the corresponding Fourier transform variable, which also is a function of time. While it is not possible to compute the average in (1) in the general case analytically, there exists an analytical expression for a Gaussian $u(t)$ [9]. Since the Lagrangian increment statistics are highly non-Gaussian for small time lags, Gaussian characteristic functionals cannot be used directly. Therefore, we consider an ensemble of Gaussian trajectories which differ with respect to the dissipation rate ε. The underlying assumption is that the velocity statistics of particles traversing regions of varying dissipation rates in the flow field can be modeled by the velocity statistics of an ensemble of particles where each particle travels through regions with a prescribed dissipation rate and simple Gaussian statistics. The characteristic functional of an ensemble of Gaussian $u(t)$ takes the form

$$\varphi[\alpha] = \int_0^\infty \mathrm{d}\varepsilon\, P(\varepsilon)\varphi_\varepsilon^G[\alpha] = \int_0^\infty \mathrm{d}\varepsilon\, P(\varepsilon) \exp\left(-\frac{1}{2} \int_{-\infty}^{\infty} \mathrm{d}t \int_{-\infty}^{\infty} \mathrm{d}t' \alpha(t) \langle u(t)u(t')\rangle_\varepsilon \alpha(t') \right). \quad (2)$$

The angular brackets with the subscript ε indicate an average over the respective ensemble members (i.e. trajectories) with identical dissipation rates. It is assumed that the mean velocity of the respective ensemble members is zero, i.e. $\langle u(t)\rangle_\varepsilon = 0$. The characteristic functional in (2) is determined by the distribution for the dissipation rate $P(\varepsilon)$ and the velocity correlation function $\langle u(t)u(t')\rangle_\varepsilon$.

In order to ensure that the ensemble one-time velocity statistics preserve Gaussianity, we construct the ensemble such that each ensemble member has the same second moment of velocity $\langle u(t)^2\rangle$. By an appropriate choice of α, the characteristic functional (2) can be projected to yield the desired joint statistics or multipoint statistics. For example, one-time velocity statistics are derived by the choice $\alpha(t) = \alpha_1\delta(t - t_1)$ and evaluating the integrals in (2). This turns the characteristic functional into a characteristic function $\varphi(\alpha_1)$. Since the second moment of velocity is by construction independent of ε, the ensemble velocity statistics are Gaussian:

$$\varphi(\alpha_1) = \int_0^\infty \mathrm{d}\varepsilon\, P(\varepsilon) \exp\left(-\frac{1}{2} \langle u(t_1)^2\rangle_\varepsilon\, \alpha_1^2 \right) = \exp\left(-\frac{1}{2} \langle u(t_1)^2\rangle\, \alpha_1^2 \right). \quad (3)$$

In contrast, the second-order moments of acceleration and increment statistics explicitly depend on ε. With the choices $\alpha(t) = -\beta_1 \frac{\mathrm{d}}{\mathrm{d}t}\delta(t - t_1)$ for acceleration statistics and $\alpha(t) = \mu_1\delta(t - t_2) - \mu_1\delta(t - t_1)$ for increment statistics, the characteristic functional (2) becomes the characteristic function for the acceleration and the velocity increment, respectively:

$$\varphi(\beta_1) = \int_0^\infty \mathrm{d}\varepsilon\, P(\varepsilon) \exp\left(-\frac{1}{2} \left\langle \left(\frac{\mathrm{d}}{\mathrm{d}t_1} u(t_1) \right)^2 \right\rangle_\varepsilon \beta_1^2 \right), \quad (4)$$

$$\varphi(\mu_1) = \int_0^\infty \mathrm{d}\varepsilon\, P(\varepsilon) \exp\left(-\frac{1}{2} \langle v^2\rangle_\varepsilon(\tau)\, \mu_1^2 \right), \quad (5)$$

with the velocity increment $v(\tau) = u(t_1 + \tau) - u(t_1)$ and the time lag $\tau = t_2 - t_1$. Even though the respective ensemble members show Gaussian acceleration and increment statistics, the ensemble statistics can become non-Gaussian by the superposition of the Gaussian ensemble members with varying dissipation rates.

All time-lag dependence in (5) is contained in the second moment of the velocity increment, while $P(\varepsilon)$ does not depend on the time lag. This marks a difference to K62 and the multifractal approach where the distribution for the dissipation rate averaged over the time lag (K62), or the distribution for the scaling exponents (multifractal approach), explicitly depends on the time lag. More detailed information on the analytical background of superimposing Gaussian characteristic functionals can be found in [10].

3 Numerical Implementation of an Ensemble of Gaussian Trajectories

As we have seen throughout the previous section, intermittent increment statistics can be constructed by an ensemble of Gaussian characteristic functionals with a dissipation rate assigned to each ensemble member and subsequent averaging over a distribution of dissipation rates. Here, we numerically create an ensemble of Gaussian velocity trajectories $u_\varepsilon(t)$ and evaluate velocity increment PDFs directly from the trajectories as a proof of concept. By a superposition of Fourier modes with random phases, we generate approximately Gaussian trajectories, cf. [11]. Since the velocity correlation $\langle u(t)u(t')\rangle_\varepsilon$ is given by the inverse Fourier transform of the energy spectrum, we directly model the energy spectrum to determine the amplitude of the Fourier coefficients. For each ensemble member, a dissipation rate ε is drawn from a distribution $P(\varepsilon)$. In the context of the K62 theory, a log-normal distribution for the averaged dissipation rate has been proposed. Also for our model a log-normal distribution is a plausible choice for the distribution for the kinetic energy dissipation rate. For the present numerical evaluation, we chose a distribution for $\ln(\varepsilon)$ with the mean $\mu_{\ln(\varepsilon)} = -0.25$ and the standard deviation $\sigma_{\ln(\varepsilon)} = 1.3$. Consequently, the coefficients of the log-normal distribution do not depend on the time lag here. The range for ε in the numerical implementation is the interval $[0.05, 50]$ which covers 98.2% of the probability distribution. The ensemble consists of 200 members, the number of Fourier modes is $n = 500{,}000$. Throughout the numerical implementation, all variables and parameters are treated in a non-dimensionalized form. The periodic domain has a size of $T = 1\,000\pi$. Then, the velocity for a trajectory in the ensemble is given by

$$u_\varepsilon(t_l) = \sum_{k=-n/2+1}^{n/2} \sqrt{E_\varepsilon(\omega_{|k|})\frac{2\pi}{T}}\, \exp(\mathrm{i}\varphi_k)\, \exp(\mathrm{i}\omega_k t_l)\,, \tag{6}$$

with $\omega_k = \frac{2\pi k}{T}$, $t_l = l\frac{T}{n}$ and $\varphi_k = -\varphi_{-k}$ uniformly distributed in the interval $[0, 2\pi]$. In order to guarantee a zero mean velocity, φ_0 is randomly set to 0 or π. In Lagrangian turbulence it is observed that the velocity correlation function decays exponentially which corresponds to a Lorentzian spectrum, cf. [12, 13]. We model the spectrum including a viscous cut-off as

$$E_\varepsilon(\omega_k) = \frac{1}{\pi}\frac{T_L(\varepsilon)}{1+(\omega_k T_L(\varepsilon))^2}\langle u^2\rangle\, A_\varepsilon \exp\left(-\omega_k \tau_\eta(\varepsilon)\right). \tag{7}$$

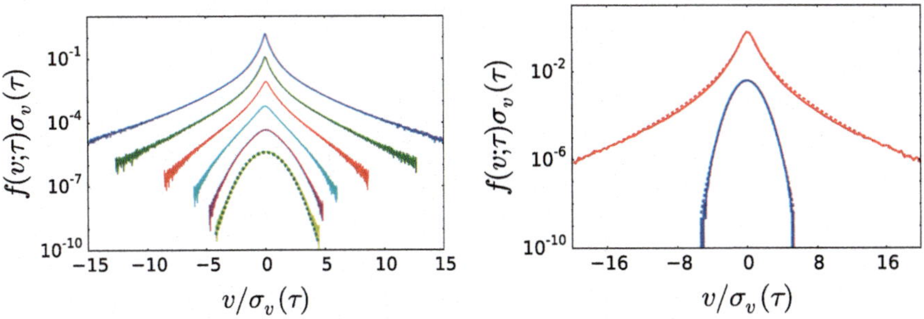

Fig. 1 *Left* Standardized PDFs for velocity increments obtained from the ensemble of Gaussian trajectories. The curves are shifted vertically for clarity. The time lags are increasing from top to bottom: {0.27, 1.07, 4.28, 17.12, 68.49, 273.96} $\tau_{\eta\langle\varepsilon\rangle}$. The Kolmogorov time scale $\tau_{\eta\langle\varepsilon\rangle}$ is determined through the expectation value of the log-normally distributed ε. The *dashed bottom curve* is a standardized Gaussian PDF as a reference which overlies the PDF for the largest time lag. *Right* Comparison of standardized velocity increment PDFs from DNS data (*solid lines*) and preliminary results of an evaluation of equation (8) (*dashed lines*) with an optimized $P(\varepsilon)$. $P(\varepsilon)$ used in this example is chosen as a log-normal distribution where the mean $\mu_{\ln(\varepsilon)}$ and the standard deviation $\sigma_{\ln(\varepsilon)}$ of $\ln(\varepsilon)$ are fitted for a small (*top curve*) and a large (*bottom curve*) time lag yielding $\mu_{\ln(\varepsilon)} = -2.8$ and $\sigma_{\ln(\varepsilon)} = 1.03$. The curves are shifted vertically for clarity. The graphs are created with matplotlib [14]

The factor A_ε is determined such that the kinetic energy $\int_0^\infty E_\varepsilon(\omega)\mathrm{d}\omega = \frac{1}{2}\langle u^2\rangle$ is kept fixed to one for each ensemble member. The choice of ε determines the integral time scale which we here define as $T_L(\varepsilon) = \frac{1}{2}\frac{\langle u^2\rangle}{\varepsilon}$ and the Kolmogorov time scale $\tau_\eta = (\nu/\varepsilon)^{1/2}$ where we choose $\nu = 0.001$.

Once the ensemble is numerically generated, arbitrary statistics can be obtained. As mentioned above, we are interested in velocity increment PDFs as an example. Equation (6) is used to compute velocity increments via $v_\varepsilon(\tau) = u_\varepsilon(t+\tau) - u_\varepsilon(t)$. We present the velocity increment PDFs in Fig. 1 on the left which shows a continuous transition from a Gaussian PDF for large time lags to a highly non-Gaussian PDF for smaller time lags. This demonstrates that it is in principle possible to generate intermittent increment statistics by an ensemble of Gaussian trajectories with varying dissipation rates.

4 Outlook: Validation with DNS Results

The numerical proof of concept in Sect. 3 shows that an ensemble of Gaussian trajectories captures intermittency qualitatively by model assumptions about the forms of the underlying spectra and the distribution $P(\varepsilon)$. The application to turbulence DNS data is work in progress. In contrast to the procedure presented in the previous section, we here take the Fourier transform of (5) as a starting point. Here, the velocity increment PDF is given by a superposition of Gaussian increment distributions which vary in terms of their variances:

$$f(v;\tau) = \int_0^\infty d\varepsilon\, P(\varepsilon) \frac{1}{\sqrt{2\pi \langle v^2\rangle_\varepsilon(\tau)}} \exp\left(-\frac{v^2}{2\langle v^2\rangle_\varepsilon(\tau)}\right). \tag{8}$$

We choose a lognormal distribution for $P(\varepsilon)$ and obtain the mean and the standard deviation of $\ln(\varepsilon)$ via an optimization procedure. The goal of the optimization procedure is to determine the parameters of $P(\varepsilon)$ in (8) for a small and a large time lag as a best fit to the respective increment distributions from DNS data.

To generate a test data set, a standard pseudo-spectral DNS at $\mathrm{Re}_\lambda = 109$ was run in a quasi-stationary regime with a resolution of $k_M\eta \approx 2$ where k_M is the maximum wavenumber and η is the Kolmogorov length. 150,000 particles were placed in this flow. The velocity of the particles was sampled along their trajectories. Lagrangian increment statistics were then estimated from the DNS data.

Figure 1 on the right shows preliminary results for the comparison of the DNS data to the superposition of Gaussian increment distributions which are weighted according to an optimized log-normal distribution $P(\varepsilon)$. The approach is in general capable of reproducing the velocity increment PDF for a small and a large time lag. We are currently extending the approach to represent Lagrangian increment statistics across the full range of temporal scales and plan to explore the model with respect to more general statistical quantities including joint statistics of velocities and acceleration.

Acknowledgements This work was supported by the Max Planck Society. We gratefully acknowledge the DNS data provided by Cristian Lalescu.

References

1. A.N. Kolmogorov, The local structure of turbulence in incompressible viscous fluid for very large Reynolds numbers. Dokl. Akad. Nauk SSSR. **30** (1941), Reprinted translation by V. Levin: Proc. R. Soc. Lond. A **434**, 9–13 (1991)
2. A.S. Monin, A.M. Yaglom, *Statistical Fluid Mechanics: Mechanics of Turbulence*. Dover Books on Physics, vol. II (Dover Publications, Mineola, 2007)
3. A.N. Kolmogorov, A refinement of previous hypotheses concerning the local structure of turbulence in a viscous incompressible fluid at high Reynolds number. J. Fluid Mech. **13**(01), 82–85 (1962)
4. A.M. Oboukhov, Some specific features of atmospheric turbulence. J. Fluid Mech. **13**(01), 77–81 (1962)
5. R. Benzi, L. Biferale, E. Calzavarini, D. Lohse, F. Toschi, Velocity-gradient statistics along particle trajectories in turbulent flows: the refined similarity hypothesis in the Lagrangian frame. Phys. Rev. E **80**(6), 066318 (2009)
6. U. Frisch, *Turbulence: The Legacy of A.N. Kolmogorov* (Cambridge University Press, Cambridge, 1995)

7. L. Chevillard, B. Castaing, A. Arneodo, E. Lévêque, J.-F. Pinton, S.G. Roux, A phenomenological theory of Eulerian and Lagrangian velocity fluctuations in turbulent flows. C. R. Phys. **13**(9–10), 899–928 (2012)
8. B. Castaing, Y. Gagne, E.J. Hopfinger, Velocity probability density functions of high Reynolds number turbulence. Physica D **46**(2), 177–200 (1990)
9. J.L. Lumley, *Stochastic Tools in Turbulence*, Dover Books on Engineering (Dover Publications, Mineola, 2007)
10. M. Wilczek, Non-Gaussianity and intermittency in an ensemble of Gaussian fields. New J. Phys. **18**, 125009 (2016)
11. J. Jiménez, Turbulent velocity fluctuations need not be Gaussian. J. Fluid Mech. **376**, 139–147 (1998)
12. S.B. Pope, *Turbulent Flows* (Cambridge University Press, Cambridge, 2000)
13. N. Mordant, P. Metz, O. Michel, J.-F. Pinton, Measurement of Lagrangian velocity in fully developed turbulence. Phys. Rev. Lett. **87**(21), 214501 (2001)
14. J.D. Hunter, Matplotlib: a 2D graphics environment. Comput. Sci. Eng. **9**(3), 90–95 (2007)

Part II
Wall Bounded Flows

Convection Velocity Variation as a Result of Amplitude Modulation Phenomena

Artur Dróżdż and Witold Elsner

Abstract The paper discusses the method of convection velocity estimation in turbulent boundary layer using its relationship with amplitude modulation mechanism. To verify this method the two-point correlation measurements using hot-wire technique was applied in strong adverse pressure gradient flow for two Reynolds numbers. Additionally, streamwise velocity profiles were measured in the same locations. It was shown that the changes in the convection velocity due to Reynolds number and pressure gradient results from amplitude modulation mechanism. The convection velocity in the strong adverse pressure gradient region can be two times higher than the mean velocity in the buffer layer.

1 Introduction

For the understanding physics of the turbulent boundary layer (TBL) the study of the convection velocity U_C of vortical structures is extremely important. It is known that the transport velocity depends on the size of the individual structure, the stage of their development and their location in the boundary layer [1]. The most common published research focuses on the study of small scale motion in zero pressure gradient (ZPG) TBL [1–3]. Recent studies of Dróżdż and Elsner [4] indicate that the U_C in ZPG flow can be estimated using cross product term $\overline{3u_L^+ u_S^{+2}}/\overline{u^{+2}}^{3/2}$ of decomposed skewness factor (S_f) calculated according to Mathis et al. [5], where subscripts L and S denote the large and the small-scale components of the streamwise velocity fluctuations u, respectively. This term is also alternative measure of amplitude modulation [5] resulting from the large-scale motion (LSM). As the LSM becomes increasingly energetic at higher Reynolds numbers or with pressure gradient, their interaction with the inner small-scale motion is also enhanced [6–8]. This

A. Dróżdż (✉) · W. Elsner
Czestochowa University of Technology, al. Armii Krajowej 21,
Czestochowa, Poland
e-mail: arturdr@imc.pcz.pl

W. Elsner
e-mail: welsner@imc.pcz.pl

R. Örlü et al. (eds.), *Progress in Turbulence VII*, Springer Proceedings
in Physics 196, DOI 10.1007/978-3-319-57934-4_5

Table 1 Parameters of analyzed TBL profiles

Symbol	PG case	Re	H	u_τ [m/s]	Δx^+	Δx^+_{max}
◁	ZPG	6400	1.35	0.72	–	–
○	ZPG	10,200	1.32	0.37	–	–
△	APG	10,900	1.81	0.22	10	60
□	APG	18,100	1.62	0.43	40	240

was confirmed by the decrease of S_f in the flow subjected to favourable pressure gradient (FPG) conditions and an adequate rise in the flow subjected to adverse pressure gradient (APG) conditions, which was shown by Harun et al. [7]. Dróżdż [8] suggests that because of hardly observed high and low speed regions the production of small-scale turbulence in FPG can be considered rather as a random process. On the other hand in the APG, the LSM enhances the production of the small-scale turbulence, although only in high-speed regions. Therefore, it can be expected that small-scale structures in APG flow have higher U_C than the mean velocity [8].

In order to estimate the convection velocity U_S the following relation, based on amplitude modulation skewness factor term, was proposed [4]:

$$U_S^+ = U^+ + \frac{3\overline{u_L^+ u_S^{+2}}}{\overline{u^{+2}}^{3/2}} C^+ \tag{1}$$

where U^+ is the non-dimensional mean velocity and C^+ is the non-dimensional scale. The formula of convection velocity estimation using the proposed relation was introduced and verified in the TBL under ZPG conditions. It was concluded that the change of small-scale structures U_C can be the result of the amplitude modulation mechanism.

In the paper the verification of the proposed formula (1) using two point correlation method in strong APG flows was presented. This was done by employing two hot-wire probes separated by a given distance in the near wall region (see Table 1).

2 Test Section and Methodology

2.1 Test Section

The data comes from the experiment performed in the open circuit wind tunnel shown in Fig. 1, where the TBL was developed along the flat plate, which was 6870 mm long. The inlet rectangular channel with a length of 5.035 m located upstream the proper test section has the triangular corner inserts to control corner vortices and two pairs of suction gaps aimed to reduce boundary layers on the side walls. A slight inclination of the upper wall helped to keep zero pressure gradient (ZPG) conditions

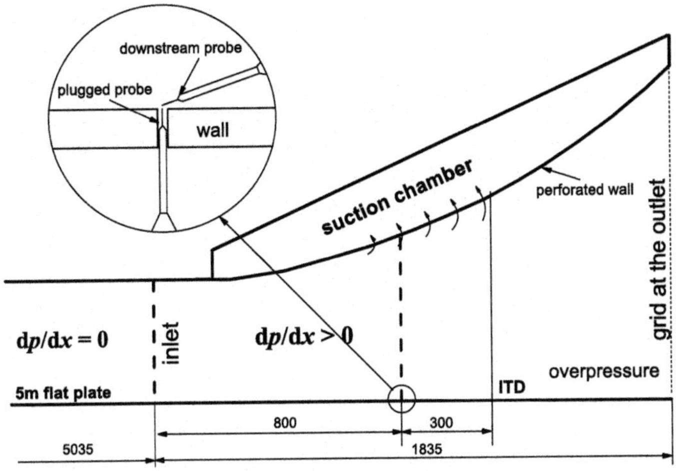

Fig. 1 Test section geometry with correlation probes setup

at the inlet. The specially designed test section located at the end of the wind-tunnel (see Fig. 1) is equipped with perforated wall. By playing with the suction flux it is possible to generate strong pressure gradient conditions leading to separation on the lower flat plate. The static pressure in the test section is increased by the throttling on the outlet of the test section.

The velocity measurements were performed with hot-wire anemometry CCC developed by Polish Academy of Science in Krakow. The analysis was conducted based on measurements of a single hot-wire probe with diameter $d = 3\,\mu$m and length $l = 0.4$ mm (modified Dantec Dynamics 55P31). The acquisition was maintained at frequency 25 kHz with minimum 30 s sampling records. For two point correlation the pair of the same probes was used.

To have the verified the reference friction velocity u_τ along the flow the fringe skin friction (FSF) technique was also applied. The facility is equipped with the computer-controlled traversing system (in streamwise x and wall-normal y direction). The traverse carriage was driven over the maximum wall displacement of 180 mm by a servo motor with the step equals 0.01 mm and uncertainty of the drive step equals 0.001 mm. In the streamwise direction the drive step was equal 0.375 mm with the uncertainty of the drive step equals 0.0375 mm.

2.2 *Convection Velocity Estimation*

The two-point correlation method employing two single hot-wire probes was used in the present work. During the measurements the first probe was plugged in the wall. The other probe was situated above the first one and shifted downstream the flow (see Fig. 1). The averaged value of U_C was calculated from six streamwise probes distances Δx on to the maximum probe distance Δx_{max} shown in Table 1. The output voltage from two hot-wires were sampled simultaneously and the time shift was obtained from conditionally averaged velocity signals processed using Wavelet Transform Analysis (WTA). Detection of the most dominant structure in the velocity signal (setting properly time scale of the wavelet) gives the good representation of the mean convection velocity. The analysis of velocity signal measured by downstream traversing probe using first derivative of Gaussian function was performed. In order to detect the accelerations and decelerations events in the signal the threshold level was applied on the transformated signal. Local extrema of the transform was used as detection time, while the maximum or minimum of wavelet transform was the criterion splitting the rapid acceleration or the rapid deceleration detections. The scale of the wavelet a was related to the scale of the dominant structure for which the maximum number of detected events N occurs. The criterion of detection threshold value applied on the wavelet transform was varied in order to obtain the high number of detections ($N > 4000$). The high number of detections ensured the smooth phase-averaged waveforms captured by the traversing probe which was the final result of the procedure. It was averaged on the time detection of the acceleration (+) and deceleration (−) events detected in velocity signal from downstream probe using the following formula:

$$\langle u(\tau)\rangle^{\pm} = \frac{1}{N^{\pm}} \sum_{i=1}^{N^{\pm}} u(t_i^{\pm} + \tau) \tag{2}$$

where t is the detection time for ith detection, while τ is the phase time.

The $\Delta\tau$ shifts of phase-averaged events on stationary probe for consecutive Δx shifts of traversing probe was used to calculate U_C. The advantage of the method is the undisturbed measurements on both probes since the traversing probe was always downstream the plugged probe, which was also very close to the wall. Because the number of accelerations and deceleration detections was different the weighted averaging, depending on the number of positive and negative events, was introduce in order to calculate mean convection velocity value.

3 Results

The convection velocity obtained from two-point correlation measurements was used to verify the U_S calculated using cross-product term of the skewness factor, $(\overline{3u_L^+ u_S^{+2}}/\overline{u^{+2}}^{3/2})$, where the cut-off timescale separating large- and small-scale signals was set on 200 viscous units. In order to verify the universality of the constant $C^+ = 16.34$ two Reynolds number in ZPG and in strong APG conditions were considered (Table 1). The pressure gradient parameter $\beta = -\frac{\delta^* U_\infty}{u_\tau^2}\frac{dU_\infty}{dx} = 17$, where δ^* is displacement thickness, and U_∞ is free stream velocity.

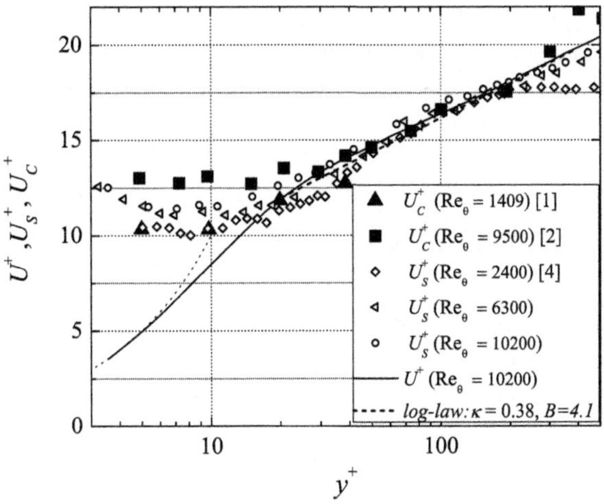

Fig. 2 Estimated convection velocities U_S and literature data values of U_C for ZPG conditions for analyzed cases (Table 1)

Figure 2 presents the estimated profiles of U_C (open points) using (1) in ZPG flow. The profiles were compared to mean velocity profile (thin black line) and to the data of Krogstad et al. [1] and Österlund [2]. It can be noticed that estimated U_C profiles are in satisfactory agreement with the literature data. The dependence of U_C^+ with Reynolds number is also observed as the amplitude modulation increases with Reynolds number.

The estimated convection velocity for APG were compared with the measured convection velocity using two-point correlation method and shown in Fig. 3. The last data are shown with the error bars related to the uncertainty of the streamwise traversing system. Additionally, the mean velocity profiles (black line) and log-law profile (dotted line) were shown. As can be seen results agree well with the profiles

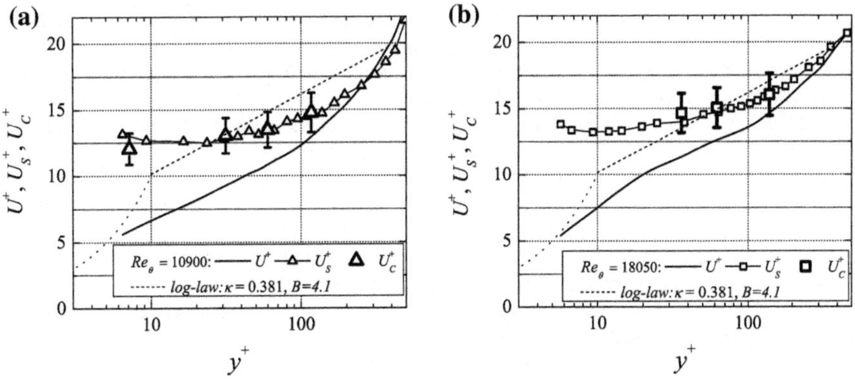

Fig. 3 Estimated convection velocities U_S and measured U_C for APG: $Re_\tau \approx 10{,}900$ (**a**), $Re_\tau \approx 18{,}100$ (**b**)

obtained according to (1). The convection velocity in the strong APG region can be two times higher than the mean velocity in the buffer layer. It can be also noticed that with the increase of Reynolds number the U_C distribution is similar to the one from ZPG conditions.

4 Conclusions

The convection velocity estimation based on the measure of amplitude modulation was verified with the convection velocity obtained from two-point correlation in the APG conditions. It was shown that the changes in the convection velocity due to Reynolds number or pressure gradient result from amplitude modulation mechanism. Distributions of the convection velocity based on the measure of amplitude modulation for both Reynolds numbers are in satisfactory agreement with the convection velocity based on two-point correlation, which confirms the correctness of C^+ value estimated for ZPG conditions. The convection velocity in the strong APG region can be two times higher than the mean velocity in the buffer layer.

Acknowledgements The investigation was supported by National Science Centre under Grant no. DEC-2012/07/B /ST8/03791 and statutory funds: BS/MN 1-103-305/2015/P and BS/MN 1-103-305/2016/P.

References

1. P.-A. Krogstad, J.H. Kaspersen, S. Rimestad, Convection velocities in a turbulent boundary layer. Phys. Fluids **10**(4), 949–957 (1998)
2. J.M. Österlund, Flow structures in zero pressure-gradient turbulent boundary layers at high Reynolds numbers. Eur. J. Mech. B/Fluids **22**(4), 379–390 (2003).
3. C. Atkinson, N.A. Buchmann, J. Soria, An experimental investigation of turbulent convection velocities in a turbulent boundary layer. Flow Turbul. Combust. **94**, 79–95 (2015).
4. A. Dróżdż, W. Elsner, Amplitude modulation and its relation to streamwise convection velocity. Int. J. Heat Fluid Flow. **63**, 67–74 (2017)
5. R. Mathis, I. Marusic, N. Hutchins, K.R. Sreenivasan, The relationship between the velocity skewness and the amplitude modulation of the small scale by the large scale in turbulent boundary layers. Phys. Fluids **23**(12), 121702 (2011).
6. R. Mathis, N. Hutchins, I. Marusic, Large-scale amplitude modulation of the small-scale structures in turbulent boundary layers. J. Fluid Mech. **628**, 311 (2009).
7. Z. Harun, I. Marusic, J. P. Monty, R. Mathis, Effects of pressure gradient on higher order statistics in turbulent boundary layers, in *Turbulence, Heat and Mass Transfer*, vol. 7 (2012) pp. 1–12
8. A. Dróżdż, Influence of pressure gradient on streamwise skewness factor in turbulent boundary layer. J. Phys.: Conf. Ser. **530**, 012061 (2014).

On the Turbulent Boundary Layer with Wall Suction

Marco Ferro, Bengt E.G. Fallenius and Jens H.M. Fransson

Abstract An experimental analysis of the turbulent boundary layer subject to wall-normal suction is carried out on a 6.4 m long perforated plate by means of hot-wire anemometry. For this type of flow the scaling of the mean velocity and of the other statistical quantities remains an open question and the amount of experimental data available, especially regarding the fluctuating velocity components, is scarce. The longer streamwise length of the present experimental apparatus compared to the one used in the previous studies allows us to better investigate the development of the boundary layer: it is shown that a turbulent asymptotic state with a constant boundary-layer thickness, analogously to what happens for the laminar state, can be closely approached experimentally and that its mean velocity profile exhibit a clear logarithmic region.

1 Introduction

Wall-normal suction and blowing is a relative simple and very effective technique to modify the behavior of a boundary layer. Suction has been used from the earliest days of boundary-layer studies to delay boundary-layer separation [1], while blowing has been explored as a possible technique to reduce friction-drag [2] and is commonly used for cooling purposes on surfaces exposed to high temperatures such as gas-turbine blades [3]. Despite the practical interests of boundary layers with wall-normal mass transfer and the numerous investigations on the topic, disagreements on the scaling of the mean velocity profile and of the other velocity statistics persist even for the simplified case of zero pressure gradient boundary-layer flow with uniform blowing or suction. This paper focuses on the case of a two-dimensional zero-pressure gradient flat-plate boundary layer with uniform wall-normal suction applied at the

M. Ferro · B.E. Fallenius · J.H.M. Fransson (✉)
Linné Flow Centre, KTH Royal Institute of Technology,
Osquars Backe 18, 100 44 Stockholm, Sweden
e-mail: jensf@kth.se

M. Ferro
e-mail: ferro@mech.kth.se

R. Örlü et al. (eds.), *Progress in Turbulence VII*, Springer Proceedings in Physics 196, DOI 10.1007/978-3-319-57934-4_6

wall. Considering the laminar regime of such a flow, it has been experimentally shown [4] that the boundary layer reaches an asymptotic state, invariant along the streamwise direction. For this asymptotic state an analytical solution of the Navier–Stokes equations exists and takes the exponential form

$$U/U_\infty = 1 - e^{V_0 y/\nu}, \tag{1}$$

where $V_0 < 0$ is the suction velocity, y the wall-normal coordinate, U the streamwise velocity and ν the kinematic viscosity of the fluid.

The possibility of obtaining a turbulent asymptotic state with a constant boundary-layer thickness, analogously to what happens for the laminar state, requires further investigation since contradictory results can be found in the literature: Dutton [5] concluded that there is just one suction rate $\Gamma = 1000 \cdot V_0/U_\infty$ for which an asymptotic state can be observed, Black and Sarnecki [6] proposed that for every suction rate (small enough to avoid relaminarization) an asymptotic state with $Re_\theta = f(\Gamma)$ can be achieved, while Tennekes [7] suggested that no asymptotic state can be obtained for suction rates lower than a certain threshold. More recently, a study by Bobke et al. [8] presented large-eddy simulations results on turbulent asymptotic suction boundary layers and raised doubts on the possibility of obtaining the asymptotic state in a practically realizable experiment.

Different scalings of the mean velocity profile have been proposed for the turbulent boundary layer with suction. As any other turbulent boundary layer flow, the turbulent suction boundary layer can be divided in two regions, a viscous sublayer where the viscous stress are prevalent and a turbulent layer where the Reynolds stresses dominates. The asymptotic description of the viscous sublayer can be readily derived as:

$$U^+ = \frac{1}{V_0^+}\left(e^{y^+ V_0^+} - 1\right), \tag{2}$$

where the superscript + indicates normalization in viscous units. For the turbulent region, instead, two different scalings have been proposed.

A bi-logarithmic law where the streamwise velocity is proportional to the squared logarithm of the wall-normal coordinate has been derived using Prandtl's momentum transfer theory by a number of authors [6, 9–12] and more recently via analytical methods [13]. The bi-logarithmic law can be expressed in the form

$$\frac{2}{V_0^+}\left(\sqrt{1 + U^+ V_0^+} - 1\right) = \frac{1}{\kappa}\ln y^+ + B, \tag{3}$$

the L.H.S. of which is sometimes referred to as pseudo-velocity. There is no agreement on the numerical values of the parameters κ and B, which in general should be considered function of the suction velocity, nevertheless a common choice among the supporter of the bi-logarithmic scaling is to set κ to the value of the non-transpired case.

Other authors [5, 7, 8, 14] have instead proposed a logarithmic dependency of the streamwise velocity on the wall-normal coordinate, analogously to what is found for non-transpired boundary layers

$$U^+ = A \ln y^+ + B, \tag{4}$$

with the slope A and the intercept B of the line dependent on the suction level.

2 Experimental Setup and Data Reduction

The experiments were conducted in the MTL wind tunnel of the Odqvist Laboratory of KTH Royal Institute of Technology. The experimental setup consists in a 6.4 m long flat-plate, with a top surface made of titanium sheets with 60 μm laser-drilled holes with centre-to-centre distance of 0.75 mm. The flat plate is installed in the wind tunnel such that the test surface constitutes the wind-tunnel bottom floor. A bleed slot between the wind-tunnel contraction and the plate leading edge allows the development of a fresh boundary layer on the test plate. The flow is tripped with a series of V-shaped embossing tapes, and is let develop on a non-transpired surface for a certain downstream distance. After this initial length, uniform wall suction is applied along the surface in the downstream direction. The streamwise component of velocity has been measured with single wire hot-wire probes with an expected accuracy of ±1%. The suction velocity is obtained measuring the pressure drop across the sheet in combination with permeability measurement. The accuracy on the permeability calibration and hence on the suction velocity is ±2% giving an expected accuracy on the suction rate of ±3%. For the cases without suction, the friction velocity u_τ, used to normalize in viscous units, has been obtained from a fit of the near-wall mean velocity data to the composite profile proposed in [15]. For all the suction cases, instead, u_τ has been obtained from Von-Kármán momentum integral modified for mass-transfer

$$\left(\frac{u_\tau}{U_\infty}\right)^2 = \frac{C_f}{2} = \frac{\mathrm{d}\theta}{\mathrm{d}x} - \frac{V_0}{U_\infty}, \tag{5}$$

where $\mathrm{d}\theta/\mathrm{d}x$ was obtained from a second order polynomial fit through the measured momentum thicknesses. Since for the reported experiments the first term of the R.H.S. of (5) is at least one order of magnitude smaller than the second term, C_f has the same uncertainty as the suction rate.

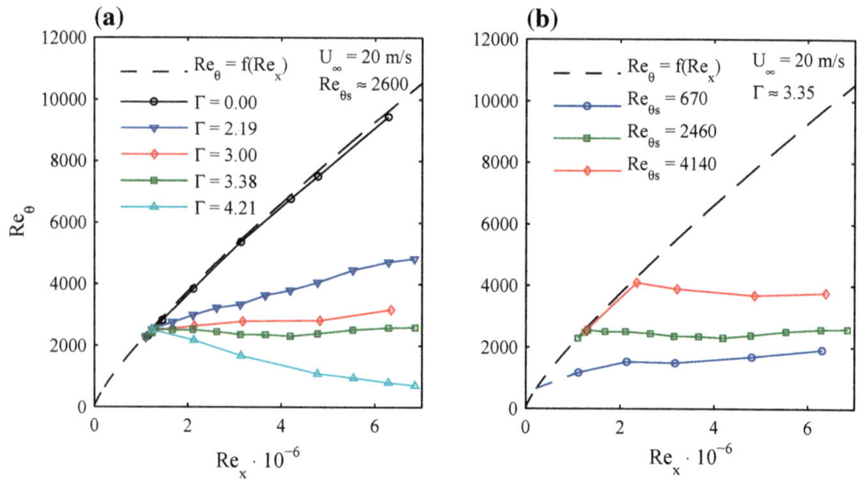

Fig. 1 Momentum-thickness Re_θ evolution along the streamwise coordinate Re_x for: **a** different suction ratio and matching $Re_{\theta s}$; **b** matching suction ratio but different $Re_{\theta s}$. *Dashed line* empirical relation $Re_\theta = f(Re_x)$ for canonical ZPG TBL from [16]

3 Results and Discussion

The evolution of the momentum thickness Re_θ along the streamwise coordinate Re_x for different suction rates is illustrated in Fig. 1a. $Re_{\theta s}$ represents the boundary-layer momentum thickness at the position where the suction starts to be applied. It should be noted that for the largest suction rate ($\Gamma = 4.21$) relaminarization is observed. For $\Gamma = 3.38$ the max-min variation around the mean of the measured momentum thicknesses is limited to $\pm 10\%$ along the whole measurement domain, reducing to $\pm 2\%$ if just the last three measurement locations are considered ($\Delta x \approx 65\,\delta_{95}$). Even though these observations suggest that a turbulent asymptotic state has been obtained, caution should be exercised: Fig. 2, showing the mean and r.m.s. streamwise velocity profiles for $\Gamma = 3.38$ at different streamwise position, illustrates that the boundary layer undergoes a slow evolution along the streamwise direction at almost constant momentum thickness. It is just towards the end of the domain that a streamwise invariant mean velocity profile is obtained.

Figure 1b shows the effect of a variation of the non-transpired entry length for constant free-stream velocity and suction rate. We notice how the boundary layer seems to approach different thicknesses for different entry lengths, as already reported in [5]. Even though a dependency of the asymptotic state might be discerned, additional measurement are required prior to any firm conclusion.

Figure 3 shows turbulent suction profiles obtained at a fixed streamwise location and different suction rates. We observe that a large portion of the profile exhibit a clear logarithmic region for the mean streamwise velocity. For $\Gamma = 3.00$ and 3.38 the boundary layer profile do not present a wake region, and the logarithmic region

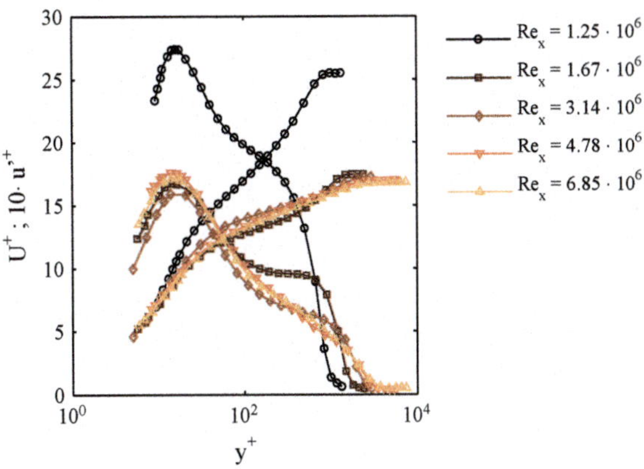

Fig. 2 Evolution of the inner-scaled mean and r.m.s. streamwise velocity profile along the streamwise direction. $\Gamma = 3.38$ and $Re_{\theta s} = 2600$. The profile at $Re_x = 1.25 \times 10^6$ is measured on the non-transpired initial length

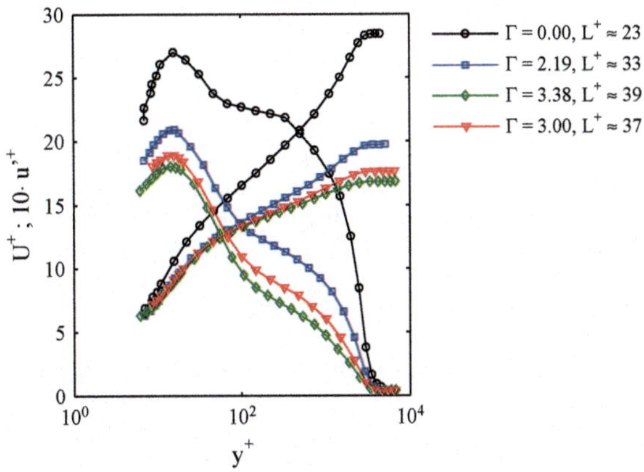

Fig. 3 Inner-scaled mean and r.m.s. streamwise velocity profile at fixed streamwise location $Re_x = 6.30 \times 10^6$ for the turbulent cases reported in Fig. 1a

appears to extend almost until the free-stream, while for $\Gamma = 2.19$ a wake region where the mean velocity departs from its logarithmic behavior can be observed. It should anyway be noticed that for $\Gamma = 2.19$ the momentum thickness is monotonically increasing in the measurement domain, i.e. the boundary layer is still far from its asymptotic state.

4 Conclusion and Outlook

Preliminary measurements on a newly built apparatus for studies on boundary layers with mass transfer suggest that a turbulent asymptotic boundary layer state can be experimentally obtained. The asymptotic profiles of mean streamwise velocity are characterized by a clear logarithmic region extending almost until the free stream, i.e. without a wake region. In order to assess the influence of the initial conditions on the asymptotic state and the dependency on the suction rate of the log-region parameters a larger dataset is required. Efforts in this sense have been initiated and new experiments are currently ongoing.

References

1. L. Prandtl: Motion of fluids with very little viscosity, in *NACA-TM 452* (1928). (Transl., original work published 1904)
2. V. Kornilov, Current state and prospects of researches on the control of turbulent boundary layer by air blowing. Prog. Aerosp. Sci. **76**, 1–23 (2015)
3. R.J. Goldstein, Film cooling, in *Advances in Heat Transfer*, ed. by T.F. Irvine, J.P. Hartnett (Elsevier, New York, 1971), pp. 321–379
4. J.M. Kay, Boundary-layer flow along a flat plate with uniform suction. Aero. Res. Council R. & M. **2628** (1948)
5. R.A. Dutton, The effects of distributed suction on the development of turbulent boundary layers. Aero. Res. Council R. & M. **3155** (1958)
6. T.J. Black, A.J. Sarnecki, W.A. Mair, The turbulent boundary layer with suction or injection. Aero. Res. Council. R. & M. **3387** (1958)
7. H. Tennekes, Similarity laws for turbulent boundary layers with suction or injection. J. Fluid Mech. **21**(4), 689–703 (1965)
8. A. Bobke, R. Örlü, P. Schlatter, Simulations of turbulent asymptotic suction boundary layers. J. Turb. **17**(2), 157–180 (2016)
9. J.H. Clarke, H.R. Menkes, P.A. Libby, A provisional analysis of turbulent boundary layers with injection. J. Aeronaut. Sci. **22**(4), 255–260 (1955)
10. H.S. Mickley, R.S. Davis, Momentum transfer for flow over a flat plate with blowing, in *NACA-TN 4017* (1957)
11. R.L. Simpson, Characteristics of turbulent boundary layers at low reynolds numbers with and without transpiration. J. Fluid Mech. **42**(4), 769–802 (1970)
12. T.N. Stevenson, A law of the wall for turbulent boundary layers with suction or injection, in *Cranfield College of Aero Report*, vol. 166 (1963)
13. I. Vigdorovich, M. Oberlack, Analytical study of turbulent poiseuille flow with wall transpiration. Phys. Fluids **20**(5), 055102 (2008)
14. P.S. Andersen, W.M. Kays, R.J. Moffat, The turbulent boundary layer on a porous plate: an experimental study of the fluid mechanics for adverse free stream pressure gradients, in *NASA-CR*, vol. 127817 (1972)
15. H.M. Nagib, K.A. Chauhan, Variations of von Kármán coefficient in canonical flows. Phys. Fluids **20**(10), 101518 (2008)
16. H.M. Nagib, K.A. Chauhan, P.A. Monkewitz, Approach to an asymptotic state for zero pressure gradient turbulent boundary layers. Philos. Trans. R. Soc. A **365**(1852), 755–770 (2007)

DNS of Couette Flows With Wall Transpiration up to $Re_\tau = 1000$

Stefanie Kraheberger, Sergio Hoyas and Martin Oberlack

Abstract We present a new set of direct numerical simulation data of turbulent plane Couette flow with constant wall-normal transpiration velocity V_0, i.e. permeable boundary conditions, such that there is blowing on the lower side and suction on the upper. Hence, there is no net change in flux to preserve periodic boundary conditions in streamwise direction. Simulations were performed at $Re_\tau = 250, 500, 1000$ with varying transpiration rates in the range of $V_0^+ \approx 0.03$–0.07. Additionally, a classical Couette flow case at $Re_\tau = 1000$ is presented for comparison. Regarding the mean velocity profile, we found a considerably extended logarithmic region with constant indicator function at $\kappa = 0.77$ as transpiration increases. Turbulent intensities are observed to decrease with increasing transpiration rate. Mean velocities and intensities collapse only in the cases where the transpiration rate is kept constant, while they are largely insensitive to friction Reynolds number variation. The statistics of these simulations can be downloaded from the webpage of the http://www.fdy.tu-darmstadt.de/fdy/fdyresearch/dns/direkte_numerische_simulation.en.jsp `Chair of Fluid Dynamics.`

1 Introduction

Direct numerical simulation (DNS) is a fundamental tool for the study of wall turbulence, and the only available one when experiments are difficult, or simply impossible,

S. Kraheberger (✉) · M. Oberlack
Chair of Fluid Dynamics, TU Darmstadt, Otto-Berndt-Str. 2,
64287 Darmstadt, Germany
e-mail: kraheberger@fdy.tu-darmstadt.de

S. Kraheberger · M. Oberlack
Graduate School of Excellence Computational Engineering (GSCE), TU Darmstadt,
Dolivostrasse 15, 64293 Darmstadt, Germany
e-mail: oberlack@fdy.tu-darmstadt.de

S. Hoyas
Instituto de Matemática Pura y Aplicada, Universitat Politècnica de València,
Camino de Vera, 46024 Valencia, Spain
e-mail: serhocal@mot.upv.es

R. Örlü et al. (eds.), *Progress in Turbulence VII*, Springer Proceedings
in Physics 196, DOI 10.1007/978-3-319-57934-4_7

to perform. Focusing on turbulent channel flow, there has been a continuous increase in Reynolds number of simulation of Poiseuille flows [4, 5]. Couette flows have been less studied, as the study of this flow is much more computationally expensive than turbulent Poiseuille flow. In the case of non-canonical boundary conditions such as wall-normal transpiration, the bibliography for channel flow is considerably reduced. For Poiseuille Transpiration Flow (PTF), the interest reader is referred to [2]. In the present work, a new set of DNS of a plane CTF has been conducted within a computational box of $L_x = 8\pi h$, $L_y = 2h$ and $L_z = 3\pi h$, with spanwise and streamwise periodicity. Details of the numerical procedure can be found in [1]. The streamwise, wall-normal, spanwise coordinates are x, y, and z and the corresponding velocity components are U, V and W or, using index notation, U_i. Statistically averaged quantities are denoted by an overbar, whereas fluctuating quantities are denoted by lowercase letters, i.e. $U = \bar{U} + u$. The flow is driven by a constant velocity of the upper wall such that we have the boundary condition $U(x, 0, z) = 0$ and $U(x, 2h, z) = U_w$. The blowing-suction process is implemented through the following boundary condition at the two walls, $V(x, 0, z) = V(x, 2h, z) = V_0$, where V_0 is the constant transpiration velocity. The nominal Reynolds numbers studied are $Re_\tau = 250$, 500 and 1000, based on the mean friction velocity u_τ and on the channel half-width h. The mean friction velocity is defined as

$$u_\tau = \sqrt{\frac{u_{\tau b}^2 + u_{\tau s}^2}{2}}, \tag{1}$$

where the local friction velocities are $u_{\tau b} = \sqrt{\nu \left|\partial_y U\right|_b}$ and $u_{\tau s} = \sqrt{\nu \left|\partial_y U\right|_s}$. Here and subsequently, subscripts b and s correspond to variables taken on the blowing and the suction side, respectively.

Table 1 summarizes the parameters of the present simulations. The wall-normal grid spacing is adjusted to keep the resolution at $\Delta y = 1.5\eta$ and approximately constant in terms of the local isotropic Kolmogorov scale $\eta = (\nu^3/\varepsilon)^{1/4}$ for every Re_τ case. In wall units, Δy^+ varies from 0.42 at the wall up to $\Delta y^+ \simeq 7.2$ at the centerline. The resolution in Fourier Space for x and z is $\Delta x^+ \sim 12.2$ and $\Delta z^+ \sim 6.13$. The case C00 is a pure turbulent Couette flow in a $16\pi h \times 2h \times 6\pi h$ computational box used as a reference case. The transpiration velocity grows for the cases C02-C20. Cases A15 and A12 were ran to study the effect of increasing Re_τ keeping the dimensionless parameters U_w/V_0 and V_0^+ approximately constant.

One of the measures used to asses that the code has run enough time to compute accurate statistics, is to compute the total shear stress, which for the CTF reads $\tau_b + V_0\bar{U} = \nu\frac{d\bar{U}}{dy} - \overline{uv}$, and, non-dimensionalized by $u_{\tau b}$, yields

$$1 + V_0^{+b}\bar{U}^{+b} - \frac{d\bar{U}^{+b}}{dy^{+b}} + \overline{uv}^{+b} = 0. \tag{2}$$

Table 1 Parameters of the simulations. Three different Reynolds numbers are given: Re_τ is based on the mean friction velocity and the channel half-width h and Re_{V_0} is based on the transpiration velocity V_0 and h. The third column, U_w/V_0, is the ratio between the velocity of the wall U_w and V_0, which defines a Reynolds number usually employed for TASBL, see [3]. Next, the dimensionless U_w and V_0 are given in terms of the wall-velocity of the pure Couette case, U_w^{C00}, and the mean friction velocity u_τ, respectively. N_x, N_y, N_z are the numbers of collocation points. The last column denotes the computational time span during which flow statistics were taken. T is the computational time spanned by those fields. Line shapes given in the second column are used to identify the cases through all the figures of the paper

Case	Line	Re_τ	Re_{V_0}	U_w/V_0	U_w/U_w^{C00}	V_0^+	N_x	N_y	N_z	U_bT/L_x
C00	—○—	1000	0	∞	1	0	6144	383	4608	6.1
C02	······	1000	32	1243	1.382	0.0324	3072	383	2304	18.7
C05	----	1000	50	685	1.907	0.0512	3072	383	2304	17.6
C10	—·—	1000	60	492	2.741	0.0630	3072	383	2304	22.0
C20	——	1000	75	395	4.402	0.0710	3072	383	2304	19.05
A15	—△—	500	37.5	400	3.342	0.0703	1536	251	1152	24.9
A12	—▽—	250	19	400	2.673	0.0695	768	251	576	60.6

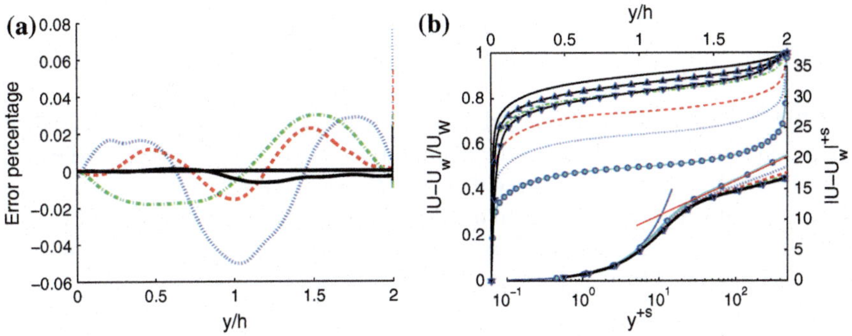

Fig. 1 Color online. Lines as in Table 1. **a** Deviation of (2) from zero given in error percentage of the computation of the transpiration CXX cases. **b** Mean velocity profile scaled in *outer* (*top-left*, U_w, h) and *inner* (*bottom-right*, u_{τ_s}, y^+) scales. In both cases, the suction wall is on the left of the figure. *Blue thin solid line* corresponds to viscous sublayer linear scaling law; *red thin solid line* represents near-wall classical logarithmic scaling law

For the most unfavorable cases, i.e. $Re_\tau = 1000$, the verification can be seen in Fig. 1a, where the deviation from 0 of the left hand side of (2) has been plotted. As some of the terms of this equation can be large, the error has been normalized by the absolute maximum value of the terms presented in (2). The deviation in all cases is less than 0.08% of the maximum value.

Another important consequence of (2) is that evaluating it at the upper wall, we obtain $u_{\tau s}^2 - u_{\tau b}^2 = V_0 U_w$, linking friction velocities with the value of the transpiration and the moving wall velocities.

2 Statistics

The mean velocity profiles may be taken from Fig. 1b. This figure shows how transpiration leads to the loss of symmetry even for small V_0^+, leading to increasingly higher mean velocity gradients at the suction wall and lower gradients at the blowing wall. In inner scaling, logarithmic behavior of the form $\bar{U} = 1/\kappa \ln y^+ + B$ can be observed, though not at the classical coefficient of $\kappa = 0.41$ and $B = 5.2$ (e.g., see [6]) but rather at $\kappa = 0.77$ and $B = 8.6$. For comparison, [3] observed $\kappa = 0.89$ and $B = 9.6$ for a TASBL at $U_\infty / V_0 = 333$.

Figure 2 shows the mean velocity profile in terms of the inverse of the Kármán constant $\frac{1}{\kappa} = y^{+s} \partial U^{+s} / \partial y^{+s}$. Apparently, with increasing transpiration rate, the region where this term is almost constant increases drastically. In fact, for the highest transpiration rate presented here, an approximately constant region between $y^{+s} = 80$ and $y^{+s} = 1000$ can be observed. In the semi-logarithmic representation Fig. 2b, it can be seen that the near-wall peak of the mean velocity in the buffer layer is reduced as transpiration is increased. Mean velocities curves of the second set of simulations (C20, A15 and A12) fall on top of each other, as the lower Reynolds number cases A12 and A15 collapse onto the $Re_\tau = 1000$ curve. One of the possible reasons for the greatly extended range of validity of the log-region might be the value of $u_{\tau s}$ which is ten times larger than the one for the classical Couette flow.

In Fig. 3, the root-mean-square velocity fluctuations $u_i'^+ = \left(\overline{u_i u_i}^{1/2}\right)^+$ and $uv'^+ = \overline{uv}^+$ are presented for the different transpiration cases to be also compared to the pure Couette case. It should be mentioned that normalization is not trivial here, since there are several velocity scales acting on the flow. Through the BC we have the external scales U_w and V_0, while internally we have the two friction velocities $u_{\tau b}$ and $u_{\tau s}$, which are all related by the global momentum balance $u_{\tau s}^2 - u_{\tau b}^2 = V_0 U_w$.

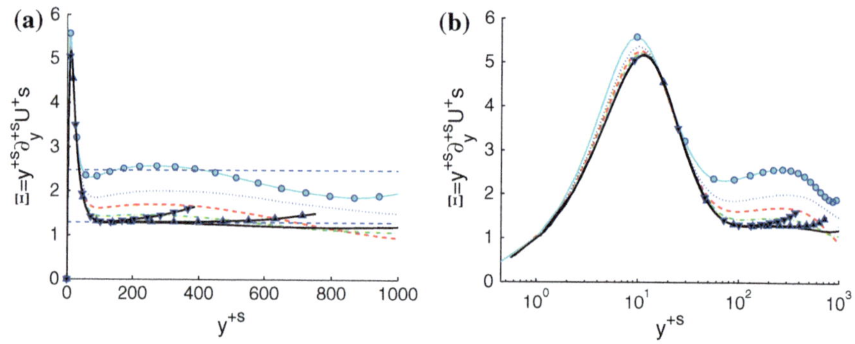

Fig. 2 Color online. Lines as in Table 1. Indicator function for the logarithmic layer, i.e. the inverse of the von-Kármán-constant scaled with $u_{\tau s}$, **a** with *dashed lines* at $\kappa = 0.41$ and $\kappa = 0.77$, **b** in semi-logarithmic plot. Suction wall is at the *left side* of the plots

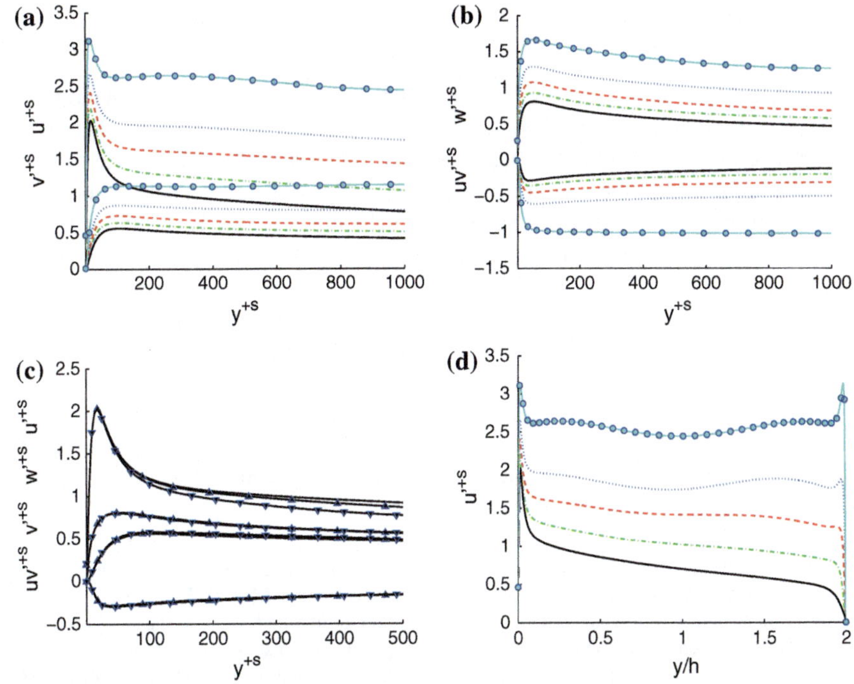

Fig. 3 Color online. Lines as in Table 1. Velocity fluctuation intensities of the CXX-cases **a** u'^{+s} and v'^{+s}; **b** w'^{+s} and $\overline{uv}^{+s}$ plotted over dimensionless distance from the suction wall in inner units ($y_s^{+s} = (2 - y/h)u_{\tau s}/\nu$). **c** Velocity fluctuation intensities of C20 and the AXX-cases from *top* to *bottom* u'^{+s}, w'^{+s}, v'^{+s} and $\overline{uv}^{+s}$ plotted over y_s^{+s}. **d** Velocity fluctuation intensity u'^{+s} of the CXX-cases plotted over dimensional distance from the suction wall in outer units

As the effect of the suction wall seems to dominate the flow rather than the blowing effect, $u_{\tau s}$ is chosen for normalization. The higher the transpiration rate V_0, the lower are the peaks observed for the fluctuations which can be taken from Fig. 3. In particular, $\overline{uv}^+$ is reduced considerably with increasing transpiration rate. For comparison, $\overline{uv}^+ \approx -1$ is observed in the channel center of classical Couette flow, see Fig. 3b and [1]. The present observations show that the momentum transfer from the streamwise to the wall-normal direction is reduced by transpiration. This kind of reduction of turbulence in the flow is somehow unexpected since we add a wall-normal flow and additionally, the local Reynolds number at the suction wall is relatively large. On the other side, we could argue that together with the removal of fluid, suction removes momentum from the flow. However, the latter argumentation fails in explaining the effect which can be seen in Fig. 3d. Apart from the previously described phenomenon that an increasing transpiration rate leads to reduced fluctuation intensities, we observe in Fig. 3d that this effect is stronger at the blowing wall, while close to the suction wall, peaks remain higher.

3 Conclusion

In conclusion, we have presented, for the first time, a set of turbulent plane Couette flow simulations extended by a wall-normal transpiration velocity. The main importance of the present paper lies in the investigation of the effects due to transpiration velocity and the search of the proper velocity scales to analyze and scale the flow. In particular, we have the external velocities U_w and V_0 extended by the local friction velocities $u_{\tau s}$ and $u_{\tau b}$, although they are all interconnected by the mean momentum (2).

The computations at the highest transpiration numbers, $V_0^+ = 0.07$, collapse in wall units for different Reynolds numbers, showing that V_0^+ is the key parameter to control the flow acting as an invariant. Regarding near-wall scaling, it is observed that at the highest Reynolds number and the highest transpiration rate, the slope constant of the log-law is equal to 0.77 representing an extremely long log region, much longer than the one that can be observed in turbulent Poiseuille or Couette flows at similar Reynolds numbers.

Acknowledgements This work was supported by the German Science Foundation (DFG) under the Grant Number OB96/39-1. SH was partially supported by project ENE2015-71333-R. The work of SK is supported by the 'Excellence Initiative' of the German Federal and State Governments and the Graduate School of Computational Engineering at TU Darmstadt. The computations of the new simulations were made possible by a generous grant of computing time from the SuperMUC Petascale System at the Leibniz Supercomputing Centre (LRZ) under project-ID pr92la.

References

1. V. Avsarkisov, S. Hoyas, M. Oberlack, J.P. García-Galache, Turbulent plane Couette flow at moderately high reynolds number. J. Fluid Mech. **751**, R1 (2014)
2. V. Avsarkisov, M. Oberlack, S. Hoyas, New scaling laws for turbulent Poiseuille flow with wall transpiration. J. Fluid Mech. **746**, 99–122 (2014)
3. A. Bobke, R. Örlü, P. Schlatter, Simulations of turbulent asymptotic suction boundary layers. J. Turbulence **17**(2), 157–180 (2015)
4. S. Hoyas, J. Jimenez, Scaling of the velocity fluctuations in turbulent channels up to $Re_\tau = 2003$. Phys. Fluids **18**(1), 011702 (2006)
5. J. Kim, P. Moin, R. Moser, Turbulence statistics in fully developed channels flows at low Reynolds numbers. J. Fluid Mech. **320**, 259–285 (1987)
6. S.B. Pope, *Turbulent Flows*. (Cambridge University Press, Cambridge, 2010)

Flow Structures and Momentum Transport in Turbulent Rotating Plane Couette Flow

Takuya Kawata and P. Henrik Alfredsson

Abstract In turbulent plane Couette flow subject to spanwise anti-cyclonic system rotation there exist both large-scale roll-cell structures that arises due to the Coriolis instability and smaller-scale turbulence. The transport equations of the large-scale and small-scale parts of the Reynolds stress are derived and we decompose the flow field into its large- and small-scale parts using spatial low-pass filtering in order to closely investigate the interaction between the roll cells and the small-scale turbulence.

1 Introduction

Turbulent RPCF (rotating plane Couette flow shown in Fig. 1a) under anti-cyclonic rotation may become unstable due to the Coriolis force [3] where the instability gives rise to streamwise-elongated roll-cell structures that take various forms depending on the Reynolds number $Re = U_{\mathrm{w}}h/\nu$ and the rotation number $Ro = 2\Omega_z h/U_{\mathrm{w}}$ [5]. Here ν is the kinematic viscosity of the fluid and U_{w}, h, and Ω_z are defined in Fig. 1a. In our earlier experimental investigations [2] we have observed that at high enough Re and a low rotation rate around $Ro \approx 0.02$ the momentum transport is especially enhanced; the wall shear stress increases at $Ro \approx 0.02$ and the mean velocity gradient becomes negative at the channel centre. In this work we further analyse these experimental data with a particular interest in the interaction between large- and small-scale structures. By a spatial filtering the velocity field is decomposed into the large- and small-scale parts, and the contributions to the Reynolds stress by each component and the inter-component transport are investigated.

T. Kawata · P.H. Alfredsson (✉)
Linné Flow Centre, KTH Royal Institute of Technology, 100-44 Stockholm, Sweden
e-mail: hal@mech.kth.se

T. Kawata
e-mail: kawata@mech.kth.se

R. Örlü et al. (eds.), *Progress in Turbulence VII*, Springer Proceedings
in Physics 196, DOI 10.1007/978-3-319-57934-4_8

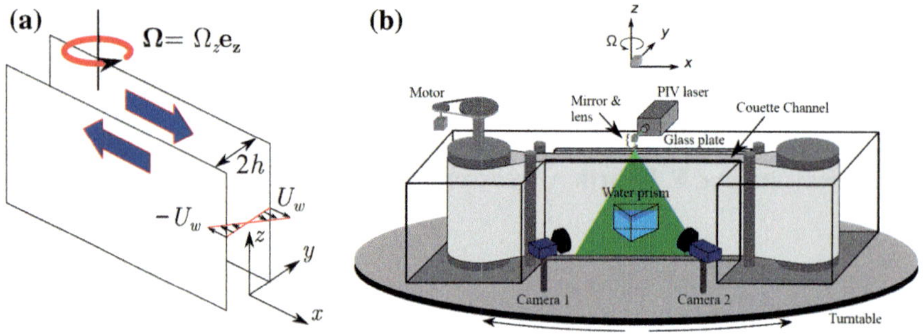

Fig. 1 **a** Definition of rotating plane Couette flow (reproduced with permission from [4]) and **b** the experimental apparatus (reproduced with permission from [1])

2 Experimental Data Sets and Analysis

The RPFC flow apparatus at KTH Mechanics, schematically shown in Fig. 1b, was used for the experiments, where the velocity field was measured with stereoscopic particle-image velocimetry (stereo PIV) in the xz-plane at ten different y-positions in the range $-0.8 < y/h < 0.4$. The Re–Ro range was $500 \leq Re \leq 2000$, $0 \leq ReRo \leq 100$. For further details the reader is referred to [2].

In the present analysis we first apply the Reynolds decomposition to the instantaneous flow field, i.e., $u_i = U_i + u'_i$ where $U_i = \overline{u_i}$ and $\overline{\alpha}$ indicates the averaged value over the xz-plane and in time of an arbitrary quantity α. The fluctuating component u'_i is further decomposed into a spatially low-pass-filtered part $\langle u'_i \rangle$ and the rest: $u''_i = u'_i - \langle u'_i \rangle$. Here we obtain $\langle u'_i \rangle$ from a spatial low-pass filter in the x- and z-directions with the same cutoff wavenumber k_{cutoff} in both directions. Since there is no overlapping wavenumber range between the filtered quantities and the rest, their correlation is zero, i.e. $\overline{\langle \alpha' \rangle \alpha''} = 0$. Hence, the Reynolds stress $\overline{u'_i u'_j}$ is simply decomposed as:

$$\overline{u'_i u'_j} = \overline{\langle u'_i \rangle \langle u'_j \rangle} + \overline{u''_i u''_j}. \tag{1}$$

The transport equation of each $\overline{\langle u'_i \rangle \langle u'_j \rangle}$ and $\overline{u''_i u''_j}$ can be derived in a similar manner as the transport equation of the 'full' Reynolds stress $\overline{u'_i u'_j}$ [5], to become

$$\left(\frac{\partial}{\partial t} + U_k \frac{\partial}{\partial x_k}\right) \overline{\langle u'_i \rangle \langle u'_j \rangle} = P^L_{ij} + G^L_{ij} + D^{t,L}_{ij} + D^{\nu,L}_{ij} + \phi^L_{ij} - \varepsilon^L_{ij} - Tr_{ij}, \tag{2}$$

$$\left(\frac{\partial}{\partial t} + U_k \frac{\partial}{\partial x_k}\right) \overline{u''_i u''_j} = P^S_{ij} + G^S_{ij} + D^{t,S}_{ij} + D^{\nu,S}_{ij} + \phi^S_{ij} - \varepsilon^S_{ij} + Tr_{ij}. \tag{3}$$

Here P_{ij}, G_{ij}, D^{ν}_{ij}, ϕ_{ij}, and ϵ_{ij} are the production, the Coriolis force term, the viscous diffusion, the velocity pressure-gradient correlation term, and the viscous dissipation with the superscripts L and S indicating their large- and small-scale parts, respectively. $D_{ij}^{t,L}$ and $D_{ij}^{t,S}$ are the large-scale and small-scale part of the turbulent diffusion term:

$$D_{ij}^{t,L} = -\frac{\partial}{\partial x_k}\left(\overline{\langle u_i'\rangle\langle u_j'\rangle\langle u_k'\rangle} + \overline{\langle u_i'\rangle\langle u_j'\rangle u_k''} + \overline{u_i''\langle u_j'\rangle u_k''} + \overline{\langle u_i'\rangle u_j'' u_k''}\right), \quad (4)$$

$$D_{ij}^{t,S} = -\frac{\partial}{\partial x_k}\left(\overline{u_i'' u_j'' u_k''} + \overline{u_i'' u_j''\langle u_k'\rangle} + \overline{\langle u_i'\rangle u_j''\langle u_k'\rangle} + \overline{u_i''\langle u_j'\rangle\langle u_k'\rangle}\right), \quad (5)$$

and Tr_{ij} is an additional term that arises due to decomposing the fluctuating velocity:

$$Tr_{ij} = \underbrace{-\overline{u_i'' u_k''\frac{\partial\langle u_j'\rangle}{\partial x_k}} - \overline{u_j'' u_k''\frac{\partial\langle u_i'\rangle}{\partial x_k}}}_{Tr_{ij}^{\mathrm{I}}} \underbrace{+ \overline{\langle u_i'\rangle\langle u_k'\rangle\frac{\partial u_j''}{\partial x_k}} + \overline{\langle u_j'\rangle\langle u_k'\rangle\frac{\partial u_i''}{\partial x_k}}}_{Tr_{ij}^{\mathrm{II}}}. \quad (6)$$

The sum of (2) and (3) yields the full Reynolds stress transport equation, and the interesting feature of these equations is that Tr_{ij} appears in both of them with different sign, which indicates that Tr_{ij} represents a transfer between the large- and small-scale parts of the Reynolds stress. Furthermore, the first two terms of Tr_{ij}, denoted as Tr_{ij}^{I}, have the same sign as the production of $\overline{u_i'' u_j''}$ whereas the other terms, denoted as Tr_{ij}^{II}, are of the same sign as the $\overline{\langle u_i'\rangle\langle u_j'\rangle}$ production. This can be interpreted that Tr_{ij}^{I} represents the energy transfer from the large- to small-scale structures, while Tr_{ij}^{II} is transport in the other direction. In the next section the spatial low-pass filtering is applied to experimental data, with focus on the energy transfer term Tr_{ij}.

3 Results and Discussion

In order to investigate over which wavenumbers the energy transfer mainly takes place, the scale interaction term of the turbulent kinetic energy was evaluated as $Tr_{k_t} = \frac{1}{2}Tr_{ii}$ using various cutoff wavenumbers k_{cutoff} and are presented in Fig. 2. The evaluated Tr_{k_t} is averaged across the channel (we used the eight measurement planes in the range $-0.8 < y/h \leq 0$ for averaging) and scaled by u_τ^3, and the black dashed lines show the cutoff wavenumber k_{cutoff} at which the integrated Tr_{k_t} takes the maximum value at each Ro. It should be noted that Tr_{k_t} contains 18 terms (see 6), but those associated with $\partial/\partial y$ cannot be evaluated based on the current stereo-PIV data, except the terms with $\partial\langle v'\rangle/\partial y$ and $\partial v''/\partial y$, which were obtained via the continuity equation.

As shown in Fig. 2, at high Re significant positive energy transfer is indicated roughly between $k_{\mathrm{cutoff}} = 0.25$ and 0.5 for $Ro \geq 0.02$, and this length scale

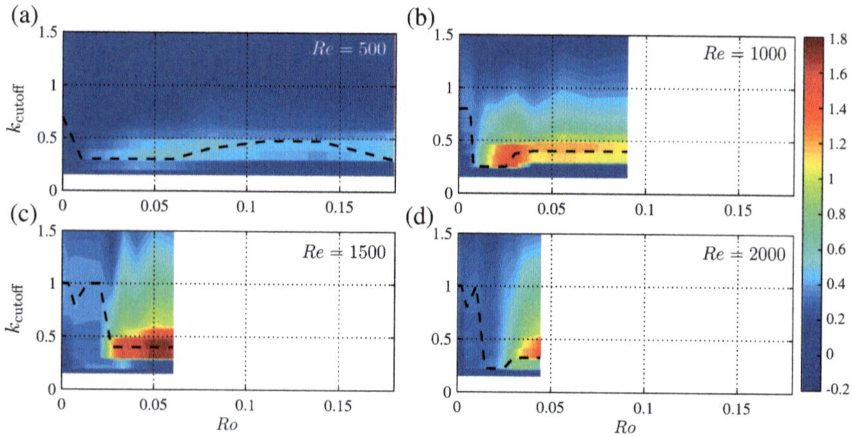

Fig. 2 The scale interaction term for the turbulent kinetic energy Tr_{k_t} evaluated with various values of cutoff wavenumbers and averaged across the channel. The values of k_{cutoff} and Tr_{k_t} values are scaled by h and u_τ^3, respectively, and the *black dashed line* in each *figure* indicates the location of the maximum value of Tr_{k_t} at each Ro

corresponds to the spanwise spacing of the roll cells. It is also seen from the dashed lines that at rotation rates close to $Ro = 0$ the maximum of the integrated Tr_{k_t} is located around $k_{\mathrm{cutoff}} \approx 1$, i.e., the scale of the half channel width. In the following analyses the spatial low-pass filtering is applied with those cutoff wavenumbers presented by the black dashed line.

Figure 3 presents a snapshot of instantaneous velocity fields at the channel centre at $Re = 2000$ for $Ro = 0$ and $Ro = 0.02$, as examples of the decomposition. At $Ro = 0$ one can see somewhat coherent and streamwise-elongated structure in the 'raw' velocity field, and the filtering with cutoff wavenumber of $k_{\mathrm{cutoff}} = 1$ separates this structure from much smaller-scale structure which looks random. Here 80% of energy is found below $k_{\mathrm{cutoff}} = 1$. In the case of higher rotation rate $Ro = 0.02$, while the cutoff wavelength is smaller ($k_{\mathrm{cutoff}} = 0.225$), the large-scale part of the flow field still retains more than 50 % of the total turbulence kinetic energy and shows a wavy-type roll-cell structure as presented in Fig. 3b.

In Fig. 4 the scale interaction term Tr_{k_t} and $-Tr_{12}$ averaged over the channel are shown comparing the Tr^{I} part and the Tr^{II} part. For $Re = 1500$ and 2000 Tr_{k_t} rapidly increases with increasing Ro around $Ro = 0.02$. It is noteworthy that despite the difference in sign between Tr_k^{I} and Tr_k^{II} (see 6) both of them are always positive, indicating the energy transfer from the large scale to small scale, and the most of the contribution is from Tr_k^{II}.

On the other hand, for the interaction term of the Reynolds shear stress $-Tr_{12}$, Tr_{12}^{I} and Tr_{12}^{II} clearly show the energy transfer in different directions in contrast to Tr_{k_t}: the former is basically positive while the latter negative, and the overall net energy transfer is negative for all Re–Ro cases, which means that the energy for $-\overline{u'v'}$ is basically transferred from the small to the large scales. It is an interesting fact that

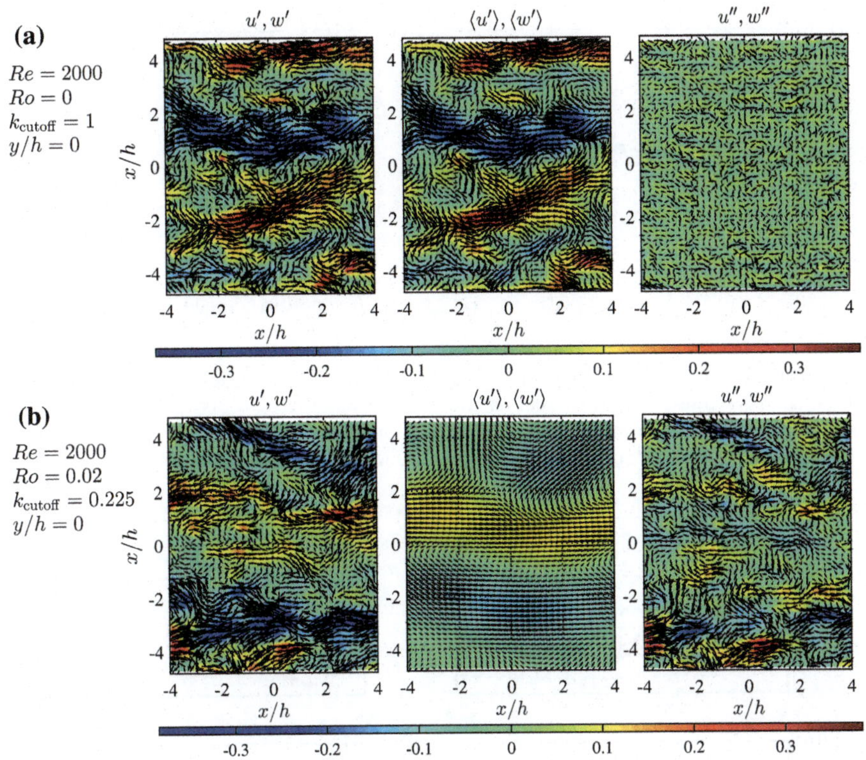

Fig. 3 Example of decomposed instantaneous velocity fields at $Re = 2000$: **a** $Ro = 0$, **b** $Ro = 0.02$. The *colour* indicates the streamwise velocity components and the *black arrows* show in-plane velocity vector pattern. The arrow length of $2.5h$ corresponds to U_w and only 1/4 of all data points are shown for the *arrows* for readability

the turbulent kinetic energy is transferred from the large to the small scales while the net energy transfer of the shear component $-\overline{u'v'}$ is in the opposite direction.

Next, the profiles of the large- and small-scale parts of the turbulent kinetic energy and the Reynolds shear stress $-\overline{u'v'}$ are shown in Fig. 5 together with their scale interaction term Tr_k and $-Tr_{12}$, comparing some different Ro cases at $Re = 2000$. One can observe in Fig. 5a that at $Ro = 0$ the large-scale part of the turbulent kinetic energy is far larger than the small-scale contribution, whereas they are comparable at higher Ro cases. As shown in Fig. 5c at high rotation rates Tr_k^{II} is significant in the near wall region.

Figure 5b shows that for the Reynolds shear stress the large-scale part $-\overline{\langle u'\rangle\langle v'\rangle}$ is dominant for all Ro cases except $Ro = 0.02$, where the wall shear stress has a narrow peak (see Fig. 15 in [2]). It is also shown that at high rotation rates the small-scale part $-\overline{u''v''}$ is partly negative in the near-wall region around $y/h \approx -0.7$ and the large-scale part $-\overline{\langle u'\rangle\langle v'\rangle}^+$ slightly exceeds unity there. Such profiles of $-\overline{\langle u'\rangle\langle v'\rangle}$ and $-\overline{u''v''}$ at high rotation rates are seen also at the other Re cases. In Fig. 5d one

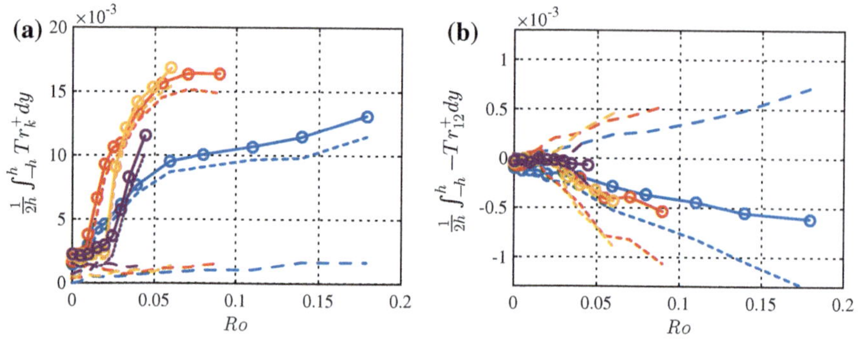

Fig. 4 Variation of the scale interaction term **a** Tr_{k_t} and **b** $-Tr_{12}$ averaged over the channel for four different Re cases: *blue* $Re = 500$; *red* $Re = 1000$; *yellow* $Re = 1500$; *purple* $Re = 2000$. The *solid lines* with *circles* indicate the 'full' values, and the *dashed* and *dotted lines* present the contribution by Tr^{I} and Tr^{II} part, respectively. The values are scaled by u_τ^4/ν

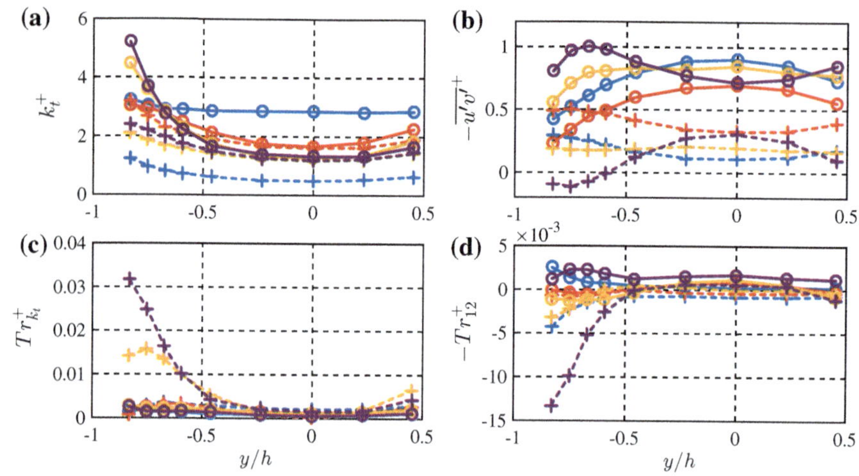

Fig. 5 Profiles of **a** the turbulent kinetic energy k_t and **b** Reynolds shear stress $-\overline{u'v'}$, and **c**, **d** their scale interaction terms scaled by the wall unit at $Re = 2000$ and four different Ro cases: *blue* $Ro = 0$; *red* $Ro = 0.02$; *yellow* $Ro = 0.03$; *purple* $Ro = 0.045$. The *solid lines* with *circles* and the *dotted lines* with + present the large-scale and the small-scale parts, respectively, in the (**a**) and (**b**), and in the (**c**) and (**d**) they show the Tr^{I} and the Tr^{II} part of each scale-interaction term, respectively

can see that the energy transfer from the small to large scales by $-Tr_{12}^{\text{II}}$ is significant in the near-wall region at high rotation rates, and this region indeed corresponds to where the $-\overline{u''v''}$ profile takes negative values. The somewhat strange profile of Reynolds shear stress at high rotation rates is a consequence of the net energy transfer from the small-scale to the large-scale motion.

References

1. T. Kawata, P.H. Alfredsson, Experiments in rotating plane Couette flow—momentum transport by roll-cell structure and zero-absolute-vorticity state. J. Fluid Mech. **791**, 191–213 (2016)
2. T. Kawata, P.H. Alfredsson, Turbulent rotating plane Couette flow: Reynolds and rotation number dependency of flow structure and momentum transport. Phys. Rev. Fluids **1**, 034402 (2016)
3. D.K. Lezius, J.P. Johnston, Roll-cell instabilities in rotating laminar and turbulent channel flows. J. Fluid Mech. **77**, 153–175 (1976)
4. A. Suryadi, A. Segalini, P.H. Alfredsson, Zero absolute vorticity: insight from experiments in rotating laminar plane Couette flow. Phys. Rev. E **89**, 033003 (2014)
5. T. Tsukahara, N. Tillmark, P.H. Alfredsson, Flow regimes in a plane Couette flow with system rotation. J. Fluid Mech. **648**, 5–33 (2010)

Towards a Physical Scale Decomposition of Mean Skin Friction Generation in the Turbulent Boundary Layer

Nicolas Renard and Sébastien Deck

Abstract A decomposition of mean skin friction generation in zero-pressure-gradient boundary layers is presented, relying on an energy budget in an absolute reference frame. It has a direct physical interpretation and emphasizes the importance of the production of turbulent kinetic energy in the logarithmic layer in mean skin friction generation at very high Reynolds number. This leads to a new approach to the scale decomposition of mean skin friction, illustrated using a Wall-Resolved LES at $Re_\theta = 13{,}000$ obtained by the ZDES technique. The role of superstructures is especially discussed.

1 Motivation and Theoretical Decomposition of Mean Skin Friction Generation into Physical Phenomena in the Boundary Layer

Because of its relation to drag, mean skin friction is essential for applied aerodynamics. Its generation is enhanced by turbulent mixing [11], leading to the well-known excess of mean skin friction of the turbulent boundary layer compared with the laminar case at the same Reynolds number. In boundary layers at high Reynolds numbers like in aerospace applications, experimental data unveiled superstructures, i.e. coherent structures of streamwise wavelength close to 5–6δ. Evaluating their contribution to mean skin friction requires to quantify the contribution of the turbulent fluctuations as a function of their wall distance and wavelength. Moreover, interpreting such an identity in terms of physical mechanisms is needed, which is all the more complicated in the zero-pressure-gradient flat plate boundary layer case as the flow is spatially developing. The FIK decomposition [6] does identify a turbulent contribution to mean skin friction, but its relation to the excess of friction is indirect because

N. Renard (✉) · S. Deck
Onera The French Aerospace Lab, 92190 Meudon, France
e-mail: nicolas.renard@onera.fr

S. Deck
e-mail: sebastien.deck@onera.fr

R. Örlü et al. (eds.), *Progress in Turbulence VII*, Springer Proceedings in Physics 196, DOI 10.1007/978-3-319-57934-4_9

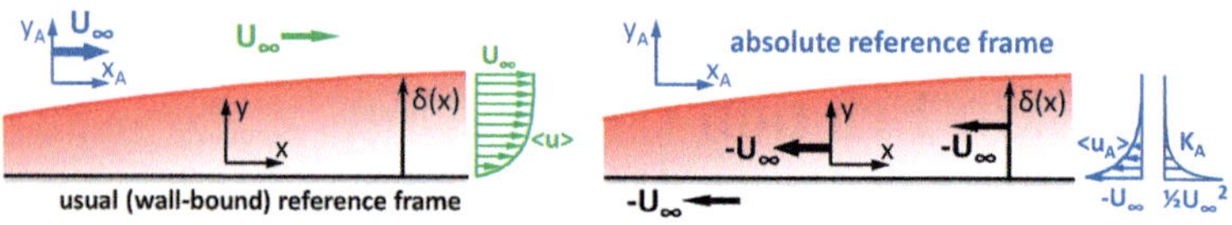

Fig. 1 Sketch of the usual and absolute frames of reference

of the spatial growth of the boundary layer [3], and the physical interpretation is not straightforward [13].

The present study is based on another decomposition of mean skin friction first introduced in [13] (where its detailed derivation is presented). This identity relies on an energy budget in a reference frame where the power of mean skin friction is non-zero (contrary to the usual wall-bound reference frame where skin friction is missing in the energy budgets because the wall is not moving). The chosen 'absolute' reference frame is associated to the outer fluid undisturbed by the wall (Fig. 1), so that the wall is moving at $-U_\infty$ along x in the absolute frame. The budget of mean streamwise kinetic energy in the absolute frame $K_A = 1/2\,\langle u_A\rangle^2$ (where $u_A = u - U_\infty$ is the streamwise velocity in the absolute frame) reads, assuming the boundary layer hypothesis in the incompressible zero-pressure-gradient flat-plate boundary layer:

$$\underbrace{\frac{\overline{D}K_A}{\overline{D}t}}_{\substack{\text{variation of } K_A\\ \text{with boundary}\\ \text{layer growth}\\ \hookrightarrow C_{f,c}(2)}} = \underbrace{\langle u_A\rangle \frac{\partial}{\partial y_A}\left(\frac{\tau}{\rho}\right)}_{\substack{\text{power of viscous}\\ \text{and turbulent efforts}\\ \hookrightarrow C_{f,c}(2)}} = \underbrace{\frac{\partial}{\partial y_A}\left(\langle u_A\rangle \frac{\tau}{\rho}\right)}_{\substack{\text{viscous and turbulent}\\ \text{diffusion of}\\ K_A\\ \hookrightarrow C_f(2)}} \underbrace{-\nu\left(\frac{\partial \langle u_A\rangle}{\partial y_A}\right)^2}_{\substack{\text{direct dissipation}\\ \text{of } K_A\\ \hookrightarrow C_{f,a}(2)}} \underbrace{+\langle u'_A v'_A\rangle \frac{\partial \langle u_A\rangle}{\partial y_A}}_{\substack{\text{'dissipation' of } K_A\\ \text{by turbulent kinetic}\\ \text{energy production}\\ \hookrightarrow C_{f,b}(2)}} \tag{1}$$

with $\frac{\overline{D}}{\overline{D}t} = \langle u\rangle \frac{\partial}{\partial x} + \langle v\rangle \frac{\partial}{\partial y} = \frac{\partial}{\partial t_A} + \langle u_A\rangle \frac{\partial}{\partial x_A} + \langle v_A\rangle \frac{\partial}{\partial y_A}$ and $\frac{\tau}{\rho} = \nu \frac{\partial \langle u\rangle}{\partial y} - \langle u'v'\rangle = \nu \frac{\partial \langle u_A\rangle}{\partial y_A} - \langle u'_A v'_A\rangle$.

Integrating the budget of K_A over the wall distance, assuming $u = 0\,(u_A = -U_\infty)$ at the smooth wall and non-dimensionalising the result leads to the following decomposition of the mean skin friction coefficient $C_f = \nu\,(\partial\langle u\rangle/\partial y)(y=0)/\left(1/2\,U_\infty^2\right)$:

$$C_f = \underbrace{\frac{2}{U_\infty^3}\int_0^\infty \nu\left(\frac{\partial \langle u\rangle}{\partial y}\right)^2 dy}_{C_{f,a}} + \underbrace{\frac{2}{U_\infty^3}\int_0^\infty -\langle u'v'\rangle \frac{\partial \langle u\rangle}{\partial y} dy}_{C_{f,b}} + \underbrace{\frac{2}{U_\infty^3}\int_0^\infty (\langle u\rangle - U_\infty)\frac{\partial}{\partial y}\left(\frac{\tau}{\rho}\right) dy}_{C_{f,c}} \tag{2}$$

This indicates that in the absolute frame, the mean energy supplied by the wall to the fluid (represented by C_f) is dissipated into heat ($C_{f,a}$), 'dissipated' by production of TKE (turbulent kinetic energy) ($C_{f,b}$) and gained as mean streamwise kinetic energy ($C_{f,c}$). The direct contribution of turbulence, $C_{f,b}$, is not associated with irreversible entropy creation since it does not involve turbulent dissipation. Focused on TKE production instead, it represents how turbulence interacts with the mean flow.

2 Role of the Turbulent Kinetic Energy Production in the Logarithmic Layer at High Reynolds Number

The decomposition of mean skin friction (2) is evaluated in Fig. 2 using DNS and WRLES datasets [5, 14–16] and RANS simulations (from the ONERA boundary layer code CLICET [1] with Jones & Launder [7] k–ε (JL) model and Michel et al. [10] model) to assess the high Reynolds number trend. This indicates that $C_{f,b}$ (2) is dominant at very high Reynolds numbers, whereas $C_{f,c}$ (2), associated with the spatial growth of the boundary layer, is negligible for $Re_\tau \to \infty$, as demonstrated in [13] (contrary to the non-negligible third term of the FIK decomposition [6] [3]). This is best understood in Fig. 3 where the integrands of each term of the decomposition (2) are non-dimensionalised and pre-multiplied so that their semi-logarithmic plot indicates the relative contribution to C_f. Because of the Reynolds invariance in the inner and outer layers, the dominant behaviour of $C_{f,b}$ is caused by the plateau of pre-multiplied TKE production in the logarithmic layer broadening with the Reynolds number (Fig. 3, detailed in [13]). Consequently, the generation of C_f at very high Reynolds number appears to be mostly driven by TKE production in the logarithmic layer (this conclusion differs from the FIK identity but is consistent with the observed importance of the logarithmic layer at high Reynolds numbers [17]). To better understand how turbulence contributes to mean skin friction, the study should focus on $C_{f,b}$, i.e. on the total TKE production, whose decomposition is attempted in the next section to identify the role of each layer and length scale.

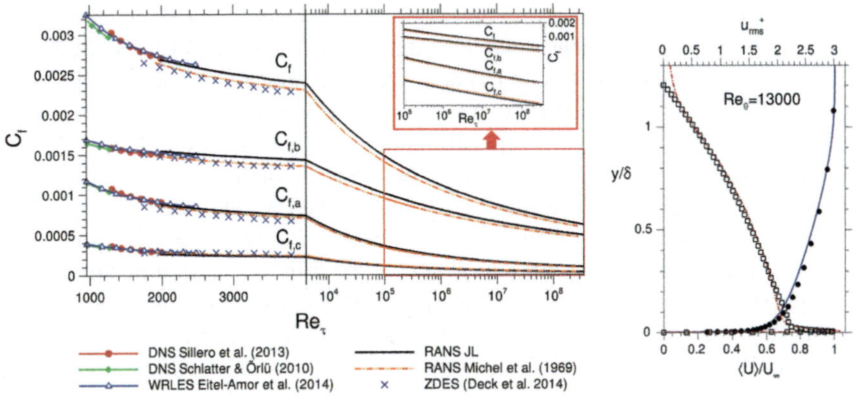

Fig. 2 Evolution of the mean skin friction decomposition (2) with the Reynolds number (*left*). ZDES profiles (see Sect. 3) at $Re_\theta = 13,000$ ($Re_\tau = 3600$): mean velocity (*solid line*) compared with experimental data by [4] (*circles*), u^+_{rms} (*dash–dotted line*) compared with the model by [9], [8] (*squares*) (*right*)

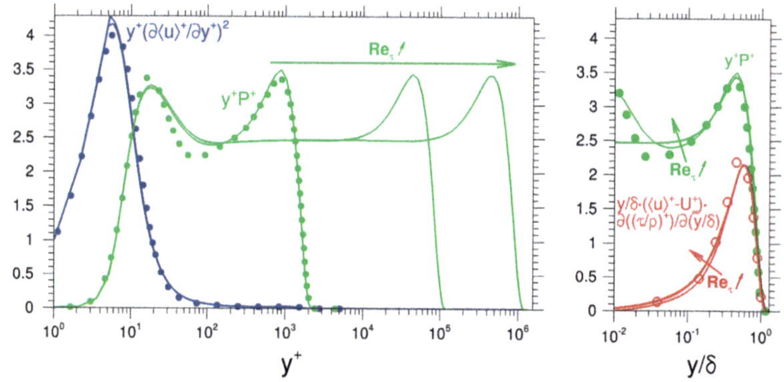

Fig. 3 Pre-multiplied integrands of the terms of the mean skin friction decomposition (2) ($y^+P^+ = -y^+\langle u'v'\rangle^+ \frac{\partial\langle u\rangle^+}{\partial y^+} = -y/\delta\,\langle u'v'\rangle^+ \frac{\partial\langle u\rangle^+}{\partial(y/\delta)}$). *Symbols* DNS [15] at $Re_\theta = 6500$ ($Re_\tau = 1989$). Lines: RANS simulations at $Re_\tau = 2000$, $Re_\tau = 10^5$ and $Re_\tau = 10^6$ (Michel et al. [10] model)

3 Scale Decomposition of Mean Skin Friction

According to the identity (2), the scale decomposition of mean skin friction is focused on decomposing the total TKE production [13]. However, the Reynolds number and scale separation are limited by the numerical cost in DNS. We consider here Re_θ = 13,000, for which no low-Mach-number boundary layer DNS dataset is available to our knowledge, and use a Wall-Resolved LES instead. This is acceptable because the study is focused on the outer layer contribution and because of the dominant role of the logarithmic layer at higher Reynolds numbers. The present WRLES is a Zonal Detached Eddy Simulation (mode III) described and validated in the outer layer in [3] (the ZDES hybrid RANS/LES technique is described in [2]). The profiles at Re_θ = 13,000 (Fig. 2) confirm the proper resolution of the outer layer. The evolution of C_f and of the terms of the identity (2) with the Reynolds number is correctly predicted as well (Fig. 2). The streamwise velocity spectra indicate that very large scale motions (much longer than 3δ) are resolved at Re_θ = 13,000, corresponding to superstructures which are missing at $Re_\theta = 5200$ in the same simulation (Fig. 4), suggesting that reaching at least the present Reynolds number is mandatory to evaluate the contribution of superstructures, which is affordable thanks to the WRLES approach. Because of the lack of scale separation at $Re_\theta = 5200$ where some inner-scaled fluctuations can share their wavelengths with outer-scaled ones, isolating the superstructures in Fourier space is not feasible. For this reason, the following illustration of the scale decomposition is performed at Re_θ = 13,000.

The scale decomposition of TKE production is obtained by estimating the co-spectrum of the Reynolds shear stress from time signals, with a spectral evaluation of the convection velocity [12]. Because a WRLES is used instead of a DNS, the emphasis is put on the outer layer and large scales, considering the cumulative

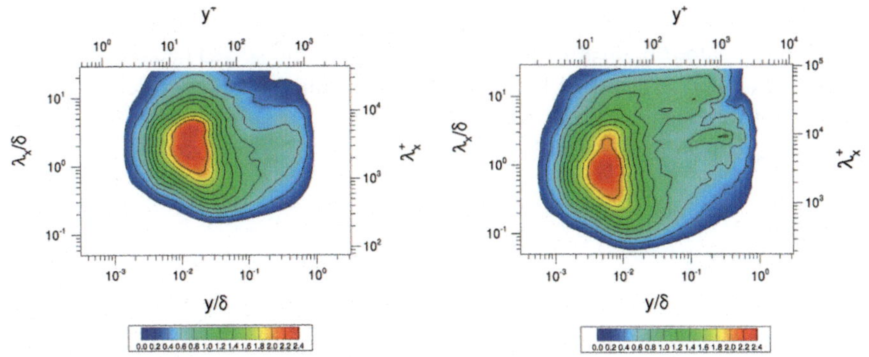

Fig. 4 Reynolds number impact on the streamwise velocity spectra $k_x G_{uu}(k_x)/u_\tau^2$ (*left* Re_θ = 5200, right: Re_θ = 13,000, reconstructed with a correlation-based convection velocity)

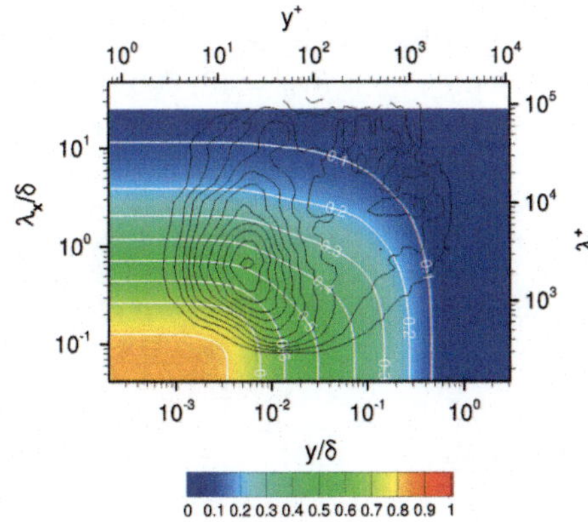

Fig. 5 Cumulative resolved contribution to TKE production $\mathscr{C}(y, \lambda_x)$ (3) at $Re_\theta = 13,000$ predicted by ZDES in WRLES mode (together with $k_x G_{uu}(k_x)/u_\tau^2$ in *black isolines*)

resolved contribution to TKE production defined as follows:

$$\mathscr{C}(y, \lambda_x) = \frac{\int_y^\infty -\left\langle u'v' \right\rangle_{\mathrm{res},[\lambda_x;+\infty[} \frac{\partial \langle u \rangle}{\partial y} dy}{\int_0^\infty -\langle u'v' \rangle \frac{\partial \langle u \rangle}{\partial y} dy} \tag{3}$$

where $-\left\langle u'v' \right\rangle_{\mathrm{res},[\lambda_x;+\infty[}$ is the Reynolds shear stress induced by the resolved fluctuations of wavelength larger than λ_x. The cumulative resolved contribution at $Re_\theta = 13,000$ is plotted in Fig. 5, showing for instance $\mathscr{C}(y^+ = 100, \lambda_x = 3\delta) \approx 0.19$, which means that the contribution of superstructures ($\lambda_x > 3\delta$) in the outer layer ($y^+ > 100$) at $Re_\theta = 13,000$ is approximately one fifth of the overall production of TKE, i.e. these fluctuations contribute at this Reynolds number approximately one fifth of $C_{f,b}$ (2), which is the dominant term in the decomposition of mean skin friction when $Re_\tau \to \infty$. However, the logarithmic layer still has a moderate extent at $Re_\theta = 13,000$. Because this layer plays an increasing role with Re_τ and since it contains the core of the superstructures, one may expect the contribution

$\mathscr{C}(y^+ = 100, \lambda_x = 3\delta)$ to increase with Re_τ, leading to a larger role of superstructures in mean skin friction at larger Reynolds numbers. Resolved datasets at higher Reynolds numbers are needed to confirm this, providing a larger scale separation than currently available from numerical simulations.

4 Outlook

A new approach to the scale decomposition of mean skin friction generation has been presented, relying on an energy budget in an absolute reference frame. It has a direct physical interpretation and emphasizes the importance of turbulent kinetic energy production in the logarithmic layer in mean skin friction at very high Reynolds number. The potential of the approach coupled to spectral analysis has been illustrated with a WRLES database obtained by the ZDES technique, showing a contribution of the superstructures in the outer layer close to one fifth of the turbulent mean skin friction term $C_{f,b}$ at $Re_\theta = 13,000$ and the need for higher-Reynolds-number wall-resolved databases to better understand the probably greater role of superstructures.

Acknowledgements The authors wish to thank all the people involved in the past and present evolution of the FLU3M code. Romain Laraufie and Pierre-Élie Weiss are warmly acknowledged for very stimulating discussions. The WRLES computation was made thanks to the HPC resources from GENCI-CINES (Project ZDESWALLTURB, Grant 2012-[c2012026817]).

References

1. B. Aupoix, Couches Limites Bidimensionnelles Compressibles. Descriptif et Mode d'emploi du Code CLICET—Version 2010. Technical Report RT 1/117015 DMAE, Onera (2010)
2. S. Deck, Recent improvements in the Zonal Detached Eddy Simulation (ZDES) formulation. Theor. Comput. Fluid. Dyn. **26**, 523–550 (2012)
3. S. Deck, N. Renard, R. Laraufie, P.E. Weiss, Large scale contribution to mean wall shear stress in high Reynolds number flat plate boundary layers up to Re_θ =13,650. J. Fluid. Mech. **743**, 202–248 (2014).
4. D.B. DeGraaff, J.K. Eaton, Reynolds number scaling of the flat plate turbulent boundary layer. J. Fluid. Mech. **422**, 319–346 (2000)
5. G. Eitel-Amor, R. Örlü, P. Schlatter, Simulation and validation of a spatially evolving turbulent boundary layer up to Re_θ = 8,300. Int. J. Heat. Fluid. Flow. **47**, 57–69 (2014)
6. K. Fukagata, K. Iwamoto, N. Kasagi, Contribution of Reynolds stress distribution to the skin friction in wall-bounded flows. Phys. Fluids. **14**(11), 73–76 (2002)
7. W. Jones, B. Launder, The prediction of laminarization with a two-equation model of turbulence. Int. J. Heat. Mass. Trans. **15**(2), 301–314 (1972)
8. I. Marusic, G. Kunkel, Streamwise turbulence intensity formulation for flat-plate boundary layers. Phys. Fluids. **15**(8), 2461–2464 (2003)
9. I. Marusic, A. Uddin, A. Perry, Similarity law for the streamwise turbulence intensity in zero-pressure-gradient turbulent boundary layers. Phys. Fluids. **9**, 3718–3726 (1997)
10. R. Michel, C. Quémard, R. Durant, Application d'un schéma de longueur de mélange à l'étude des couches limites turbulentes d'équilibre. Note Technique 154, ONERA (1969)

11. P. Orlandi, J. Jiménez, On the generation of turbulent wall friction. Phys. Fluids. **6**, 634–641 (1994)
12. N. Renard, S. Deck, On the scale-dependent turbulent convection velocity in a spatially developing flat plate turbulent boundary layer at Reynolds number $Re_\theta = 13{,}000$. J. Fluid. Mech. **775**, 105–148 (2015)
13. N. Renard, S. Deck, A theoretical decomposition of mean skin friction generation into physical phenomena across the boundary layer. J. Fluid. Mech. **790**, 339–367 (2016)
14. P. Schlatter, R. Örlü, Assessment of direct numerical simulation data of turbulent boundary layers. J. Fluid. Mech. **659**, 116–126 (2010)
15. J. Sillero, J. Jimenez, R. Moser, One-point statistics for turbulent wall-bounded flows at Reynolds numbers up to $\delta^+ \approx 2000$. Phys. Fluids. **25**, 105102 (2013)
16. J. Sillero, J. Jimenez, R. Moser, Two-point statistics for turbulent boundary layers and channels at Reynolds numbers up to $\delta^+ \approx 2000$. Phys. Fluids. **26**, 105109 (2014)
17. A.J. Smits, B.J. McKeon, I. Marusic, High-Reynolds number wall turbulence. Ann. Rev. Fluid. Mech. **43**, 353–375 (2011)

Identifying Well-Behaved Turbulent Boundary Layers

Carlos Sanmiguel Vila, Ricardo Vinuesa, Stefano Discetti, Andrea Ianiro, Philipp Schlatter and Ramis Örlü

Abstract This paper presents a study focused on the development of zero-pressure-gradient turbulent boundary layers (ZPG TBL) towards *well-behaved* conditions in the low Reynolds-number range. A new method to assess the length required for the ZPG TBL to exhibit *well-behaved* conditions is proposed. The proposed method is based on the diagnostic-plot concept (Alfredsson et al., Phys. Fluids, 23:041702, 2011), which only requires mean and turbulence intensity measurements in the outer region of the boundary layer. In contrast to the existing methods which rely on empirical skin-friction curves, shape-factor or wake-parameter, the quantities required by this method are generally much easier to measure. To test the method, the evolution of six different tripping configurations, including weak, late and strong overtripping, are studied in a wind-tunnel experiment to assess the convergence of ZPG TBLs towards *well-behaved* conditions in the momentum-thickness based Reynolds-number range $500 < Re_\theta < 4000$.

1 Introduction

The problem of establishing canonical conditions in experiments with zero-pressure-gradient (ZPG) turbulent boundary layers (TBLs) has become a relevant one since it is known that there can be important differences in quantities such as the shape factor $H = \delta^*/\theta$ (where δ^* and θ are the displacement and momentum thicknesses, respectively) or in the skin-friction coefficient c_f, due to a flawed experimental design and/or an inadequate inflow and/or development length [10]. These problems are not restricted to experimental studies since, as shown by Schlatter and Örlü [10],

C. Sanmiguel Vila · S. Discetti · A. Ianiro
Aerospace Engineering Group, Universidad Carlos III de Madrid, Leganés, Spain

R. Vinuesa · P. Schlatter · R. Örlü (✉)
Linné FLOW Centre, KTH Mechanics, SE-100 44 Stockholm, Sweden
e-mail: ramis@mech.kth.se

R. Örlü et al. (eds.), *Progress in Turbulence VII*, Springer Proceedings in Physics 196, DOI 10.1007/978-3-319-57934-4_10

numerical simulations are also affected by inflow conditions and by the tripping method. For this reason, recent studies such as the one by Marusic et al. [3] have analyzed the effect of different tripping configurations and how the flow evolves towards a canonical state. In their work, Marusic et al. [3] reported the effect of a number of tripping configurations, ranging from a weak tripping to an overstimulated case.

In the light of these findings, the need to establish criteria for the characterization of canonical conditions has emerged as an important challenge. Comparisons of experimental trends of the wake parameter Π and of H with the ones obtained from numerical integration of composite profiles are currently used as a criterion to identify *well-behaved* profiles [2], i.e., not affected by non-equilibrium effects present in the initial development stages. Moreover, Schlatter and Örlü [10] established that TBLs can be considered canonical for $Re_\theta > 2000$ if the transition is initiated prior to $Re_\theta = 300$; using this criterion, good quantitative agreement in integral quantities and higher-order moments between experiments and simulations was found. As a result of the study performed by Marusic et al. [3], it was found that the effect of the tripping mechanism was noticeable up to a streamwise distance of 2000 trip heights, although this conclusion is only applicable to their particular set-up.

All of these methods share one common characteristic: they require extensive measurements to discern whether the flow is canonical/well behaved. For this reason, here we aim at establishing a method to assess the length required for the TBL to exhibit *well-behaved* conditions, without the need to obtain velocity profiles or to measure the friction velocity at several streamwise locations. Our proposed method is based on the diagnostic-plot concept, which has been found to scale the outer layer of TBL flows irrespective of Re [1, 8]. The method will be demonstrated based on hot-wire anemometry measurements that have been performed in the Minimum Turbulence Level (MTL) wind tunnel at KTH Mechanics in which different tripping configurations in a ZPG TBL were studied.

2 Experimental Setup

The measurements were performed in the MTL closed-loop wind tunnel located at KTH Royal Institute of Technology in Stockholm. The tunnel has a 7 m long test section with a cross-sectional area of 0.8×1.2 m^2 with a streamwise velocity disturbance level lower than 0.025% of the free-stream velocity. Measurements were made in the turbulent boundary layer developing over a flat plate suspended 25 cm above the tunnel floor under a zero-pressure-gradient condition that was established through adjustment of the ceiling. All the measurements were performed at a nominal free-stream velocity of 12 m/s. Six different tripping configurations were tested using as a reference the cases studied numerically in Schlatter and Örlü [10], i.e. a combination of weak, late, and strong trippings. The different tripping configurations were placed spanning the full spanwise length of the plate, at streamwise locations in the range $75 < x$ [mm] < 230 from the leading edge (see Table 1), corresponding

Table 1 Specifications of the tripping configurations including location and respective symbol coding for symbols. The embossed (DYMO) letter 'V' points into the flow direction and has a nominal height of 0.3 mm

Tripping characteristics and location	Symbol code	Identification
DYMO 'V' @ 75 mm	Red	Weak tripping
DYMO 'V' @ 230 mm	Green	Late tripping
DYMO 'V' @ 75, 90, 110 mm and 5 mm square bar @ 85 mm	Blue	Strong overtripping
DYMO 'V' @ 90, 110 mm and 2.4 mm height turbulator	Black	Optimal 1
DYMO 'V' @ 90, 110 mm and 1.6 mm height turbulator	Magenta	Optimal 2
DYMO 'V' @ 90 mm	Cyan	Weak/late tripping

to the range $130 < Re_\theta < 260$. A set of 4 streamwise locations was selected for each tripping configuration with few additional stations to match Re_θ, covering a range of $500 < Re_\theta < 4000$. Single-point streamwise velocity measurements were performed by means of a single in-house hot-wire probe with a Platinum wire of 560 μm length and nominal diameter of 2.5 μm. These dimensions provided sufficient spatial resolution to ensure meaningful comparisons of the higher-order turbulence statistics. Care was taken to acquire sufficient measurement points within the viscous sublayer and the buffer region in order to correct for the absolute wall position and determine the friction velocity (as outlined in Örlü et al. [7]). The composite profile by Nickels [5] was used to obtain the free-stream velocity U_∞ and the 99% boundary-layer thickness δ_{99}. Reynolds numbers and integral quantities were then computed using the fitted composite profile.

3 Results and Discussion

Inner-scaled streamwise mean and variance profiles for the various trippings are shown in Fig. 1a, where it can be observed that the near-wall region quickly adapts to that of a canonical TBL [10]. On the other hand, strong variations are noticeable in the outer layer, which indicates that this part of the boundary layer requires longer development lengths to become independent of its specific tripping condition. In particular, the strong-overtripping case shows an outer peak in the fluctuation profile which is produced by the square bar used as a disturbance. In order to determine which of the TBL profiles have reached a canonical state, the Reynolds-number variation of the shape factor and the skin friction (expressed through the inner-scaled free-stream velocity U_∞^+) for all tripping configurations is depicted in Fig. 2 together with the correlations from [4, 6]. Postulating now that *well-behaved* TBL profiles should scale in the diagnostic plot (as suggested in [1, 8]), the same data set is shown in terms of the streamwise turbulence intensity u'/U versus the velocity ratio U/U_∞ in Fig. 3. Excluding, under the aforementioned premise, the profiles that do not

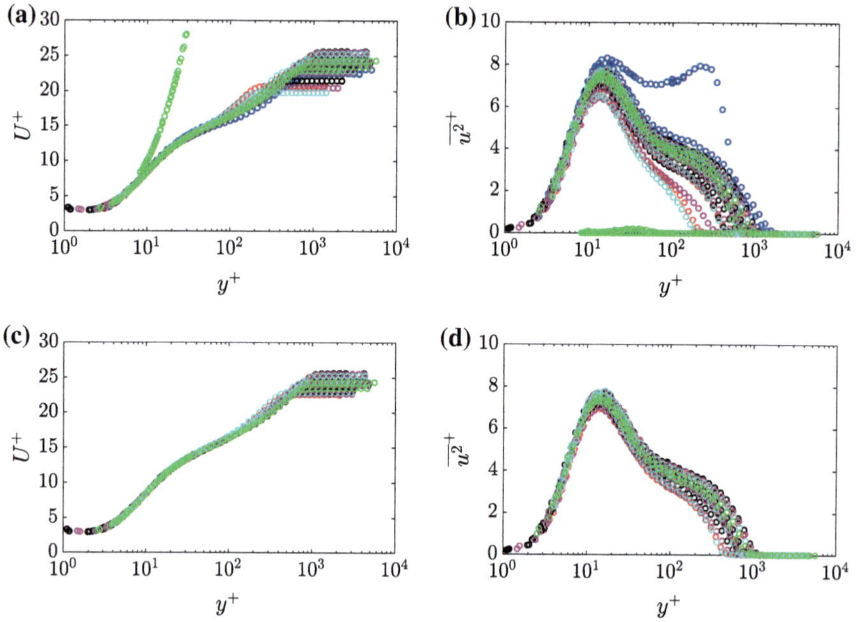

Fig. 1 Inner-scaled **a** mean and **b** variance profile for the entire data set, and **c**, **d** same quantities for the profiles that fulfill the diagnostic-plot scaling in the outer layer. See Table 1 for color code, and Fig. 3 for diagnostic-scaling plots

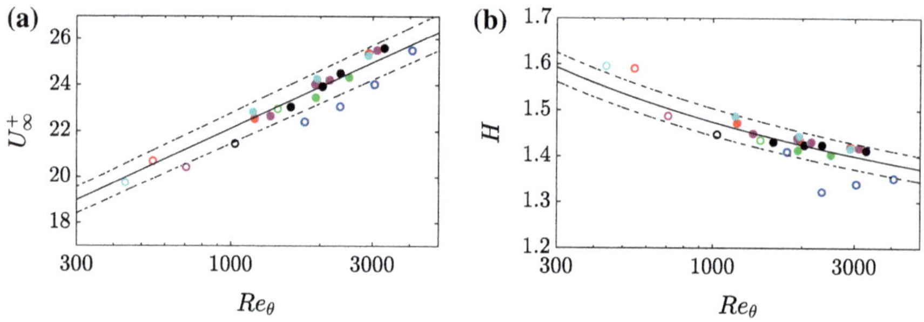

Fig. 2 **a** Shape factor H and **b** inner-scaled free-stream velocity U_∞^+ as function of Re_θ for various tripping configurations. Cases considered as *well-behaved* are further identified through filled circles. *Solid lines* represent correlations from Monkewitz et al. [4] for H and from Nagib et al. [6] for U_∞^+. *Dashed lines* are common measurement uncertainties, i.e., 3% and 2% in subplots **a** and **b**, respectively

adhere to the scaling in the outer region, especially in the region $0.7 \leq U/U_\infty \leq 0.9$, the diagnostic-plot concept provides a means to discern *well-behaved* TBLs; these profiles are indicated through filled circles in Fig. 2 and their diagnostic scaling is shown in Fig. 3b.

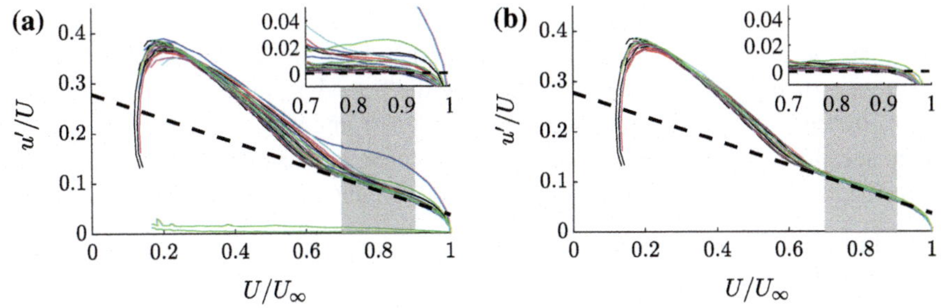

Fig. 3 Streamwise mean U and r.m.s. u' profiles plotted in diagnostic form for **a** the entire data set and **b** only the profiles that follow the diagnostic-plot scaling, i.e., those identified through *filled circles* in Fig. 2. The *dashed line* shows (*within the shaded range*) equation $u'/U = \alpha - \beta U/U_\infty$, with $\alpha = 0.280$ and $\beta = 0.245$. The *insets* show the difference between the profiles in diagnostic scaling and the diagnostic-curve fit, as a function of U/U_∞

It can be observed that the profiles that satisfy the diagnostic-plot criterion are exactly the ones that follow the reference U_∞^+ and H curves. This indicates that the diagnostic-plot criterion described in [1] is an alternative method to assess whether a particular boundary layer exhibits canonical ZPG TBL conditions. The real advantage of the proposed method is shown, that no full velocity profile measurements, integral quantities, or skin friction measurements are required. Instead a streamwise scan within the outer region of the TBL (preferably through the region of linear behavior in the diagnostic plot, i.e., the shaded area in Fig. 3) is sufficient to identify the location after which the TBL adheres to the diagnostic-plot scaling. To test this assumption, Fig. 4 shows the results of a streamwise scan in the tripping configuration weak/late tripping (see Table 1) while keeping (through an iterative procedure)

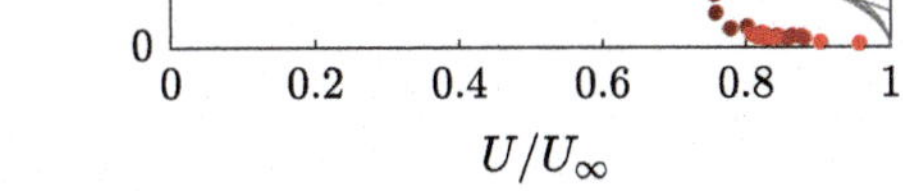

Fig. 4 A methodology based on streamwise scans and diagnostic scaling, used to predict the distance required for the ZPG TBL to exhibit *well-behaved* conditions. The method is illustrated using the tripping configuration denoted as weak/late tripping. *Solid lines* correspond to the cases of Fig. 3b. All the points are taken with an equidistant streamwise spacings of $\Delta x = 50$ mm, where *darker symbols* indicate increasing streamwise distance

the probe within the velocity range $0.7 \leq U/U_\infty \leq 0.9$. From the color-coded measurement points (from lighter to darker symbols, where darker indicates increasing streamwise distance) it can be observed how the boundary layer undergoes transition to turbulence with the overshoot in turbulence intensity and then reaches the diagnostic-plot reference curves. Hence, a simple streamwise scan easily diagnostic from which x-location on the TBL behaves in accordance with canonical ZPG TBLs.

Disclaimer Parallel to the present paper, a largely extended and more detailed study has been published by Sanmiguel Vila et al. [9].

Acknowledgements CSV acknowledges the financial support from Universidad Carlos III de Madrid within the program "Ayudas para la Movilidad del Programa Propio de Investigación". RÖ, RV and PS acknowledge the financial support from the Swedish Research Council (VR) and the Knut and Alice Wallenberg Foundation. CSV, SD and AI were partially supported by the COTURB project (Coherent Structures in Wall-bounded Turbulence), funded by the European Research Council (ERC), under grant ERC-2014.AdG-669505.

References

1. P.H. Alfredsson, A. Segalini, R. Örlü, A new scaling for the streamwise turbulence intensity in wall-bounded turbulent flows and what it tells us about the "outer" peak. Phys. Fluids **23**, 041702 (2011)
2. K.A. Chauhan, P.A. Monkewitz, H.M. Nagib, Criteria for assessing experiments in zero pressure gradient boundary layers. Fluid Dyn. Res. **41**, 021404 (2009)
3. I. Marusic, K.A. Chauhan, V. Kulandaivelu, N. Hutchins, Evolution of zero-pressure gradient boundary layers from different tripping conditions. J. Fluid Mech. **783**, 379–411 (2015)
4. P.A. Monkewitz, K.A. Chauhan, H.M. Nagib, Self-consistent high-Reynolds number asymptotics for zero-pressure-gradient turbulent boundary layers. Phys. Fluids **19**, 115101 (2007)
5. T.B. Nickels, Inner scaling for wall-bounded flows subject to large pressure gradients. J. Fluid Mech. **521**, 217–239 (2004)
6. H.M. Nagib, K.A. Chauhan, P.A. Monkewitz, Approach to an asymptotic state for zero pressure gradient turbulent boundary layers. Phil. Trans. R. Soc. **365**, 755–770 (2007)
7. R. Örlü, J.H.M. Fransson, P.H. Alfredsson, On near wall measurements of wall bounded flows – the necessity of an accurate determination of the wall position. Prog. Aero. Sci. **46**, 353–387 (2010)
8. R. Örlü, A. Segalini, J. Klewicki, P.H. Alfredsson, High-order generalisation of the diagnostic scaling for turbulent boundary layers. J. Turbul. **17**, 664–677 (2016)
9. C. Sanmiguel Vila, R. Vinuesa, S. Discetti, A. Ianiro, P. Schlatter, R. Örlü, On the identification of well-behaved turbulent boundary layers. J. Fluid Mech. **822**, 109–138 (2017).
10. P. Schlatter, R. Örlü, Turbulent boundary layers at moderate Reynolds numbers: inflow length and tripping effects. J. Fluid Mech. **710**, 5–34 (2012)

Scaling of Adverse-Pressure-Gradient Turbulent Boundary Layers in Near-Equilibrium Conditions

Ricardo Vinuesa, Alexandra Bobke, Ramis Örlü and Philipp Schlatter

Abstract Well-resolved large-eddy simulations are used to study adverse-pressure-gradient (APG) turbulent boundary layers (TBLs) under near-equilibrium conditions. In particular, we focus on two near-equilibrium cases where the power-law freestream velocity distribution is adjusted in order to produce long regions with a constant value of the Clauser pressure-gradient parameter β. In the first case we obtain an APG TBL with a constant value of $\beta \simeq 1$ over 37 average boundary-layer thicknesses, and in the second one a constant value of $\beta \simeq 2$ for around 28 average boundary-layer thicknesses. The scaling law suggested by Kitsios et al. (Int J Heat Fluid Flow 61:117–128, 2016, [10]), proposing the edge velocity and the displacement thickness as scaling parameters, was tested on the two constant-pressure-gradient parameter cases. The mean velocity and Reynolds-stress profiles were found to be dependent on the downstream development, a conclusion in agreement with classical theory.

1 Introduction

Turbulent boundary layers (TBLs) subjected to streamwise pressure gradients (PGs) are of great importance in a wide range of industrial applications, including the flow around a wing or inside a diffuser. Despite their relevance, the effects of PGs on the characteristics of wall-bounded turbulence are still elusive. Part of the reason for this lack of detailed knowledge of PG TBLs is the fact that there is no systematic approach towards characterizing the pressure gradient. Since the effect of the pressure gradient on the TBL is closely related to its streamwise development, it is important to define the concept of an *equilibrium* boundary layer: according to the strict definition by Townsend [1], this condition requires the mean flow and Reynolds-stress tensor profiles to be independent of the streamwise position x, when scaled with appropriate local velocity and length scales. As also shown by Townsend

R. Vinuesa (✉) · A. Bobke · R. Örlü · P. Schlatter
Linné FLOW Centre, KTH Mechanics, SE-100 44 Stockholm, Sweden
e-mail: rvinuesa@mech.kth.se

R. Vinuesa · A. Bobke · P. Schlatter
Swedish e-Science Research Centre (SeRC), Stockholm, Sweden

R. Örlü et al. (eds.), *Progress in Turbulence VII*, Springer Proceedings in Physics 196, DOI 10.1007/978-3-319-57934-4_11

[1] this condition is only satisfied by the sink flow, although it is possible to define a less restrictive *near-equilibrium* condition when the mean velocity defect $U_\infty - U$ is self-similar in the outer region, which in any case dominates at high Reynolds numbers. Townsend [1] and Mellor and Gibson [2] showed that these near-equilibrium conditions can be obtained when the freestream velocity is prescribed by a power law such that $U_\infty = C(x - x_0)^m$, where C is a constant, x_0 is a virtual origin and m the power-law exponent. In particular, Townsend [1] showed that m has to be in the range $-1/3 < m < 0$ in order to be in near-equilibrium conditions. An additional interesting conclusion is the fact that the widely studied zero-pressure-gradient (ZPG) TBL [3, 4], driven by a constant freestream velocity, is a particular case of the general near-equilibrium TBLs proposed by Townsend [1] and Mellor and Gibson [2] where $m = 0$. In this respect, it is relatively common in the literature to refer to "self-similar" boundary layers, where as discussed above the only case in which complete self-similarity is observed is the sink flow. For instance, Skåre and Krogstad [5] obtained an experimental APG TBL with a U_∞ distribution given by a power law, and with $m = -0.23$, which in principle would lead to near-equilibrium conditions. Nevertheless, the authors described their boundary layer as a self-similar one, therefore there is some discrepancy in the terminology and interpretation of PG TBL data. This is motivated in part by the difficulties arising from setting up canonical PG TBLs [6], as well as the impact of history effects on the local state of the TBL [7, 8], which leads to additional difficulties in the interpretation of the results.

The focus of this study is on near-equilibrium APG TBLs, and more precisely on the assessment of a particular scaling suggested by Kitsios et al. [9] for near-equilibrium conditions. To that end, we consider the Clauser pressure-gradient parameter $\beta = \delta^*/\tau_w \mathrm{d}P_\infty/\mathrm{d}x$, where δ^* is the displacement thickness, τ_w is the wall-shear stress and P_∞ is the freestream pressure, to quantify the pressure-gradient magnitude. Two well-resolved large-eddy simulations (LESs) of turbulent boundary layers in near-equilibrium, and with significant regions of the domain were β is constant, are analyzed. Note that the constant-β case can be considered as a canonical representation of an APG TBL with a particular pressure-gradient magnitude. Other APG TBLs exhibit very large variations in β, such as the TBL developing on the suction side of a wing section [10], and APG TBLs in near-equilibrium conditions (i.e., with freestream velocities given by a power-law distribution) do not necessarily lead to a region of constant β [11]. Thus, in this study we focus on the particular case of near-equilibrium APG TBLs with regions of constant β.

2 Numerical Setup

The downstream evolution of the APG TBLs was studied by means of well-resolved large-eddy simulations (LESs). The pressure gradient was imposed through the variation of the freestream velocity at the top of the domain, which was defined following the near-equilibrium definition by Townsend [1], i.e., $U_\infty(x) = C(x - x_0)^m$. We used the code SIMSON [12], which is based on a fully-spectral method with Fourier discretisation in streamwise and spanwise directions and on the Chebyshev-tau method in the wall-normal direction. Using the approximate deconvolution relaxation-term model [13] as a sub-grid scale (SGS) model as in Eitel-Amor et al. [14], the resolution was chosen as $\Delta x^+ = 21.5$, $y^+_{\max} = 16.5$ and $\Delta z^+ = 9.5$ (where x, y and z denote streamwise, wall-normal and spanwise coordinates, respectively), with 12 points below $y^+ = 10$. It is possible to obtain different near-equilibrium boundary layers by varying the virtual origin x_0 and the power-law exponent m. In the present work we focus on two configurations: a first one, where a constant value of $\beta \simeq 1$ was achieved over a streamwise distance of $37\overline{\delta}_{99}$, and a second one in which a region of constant $\beta \simeq 2$ over $28\overline{\delta}_{99}$ was obtained. Note that δ_{99} is the 99% boundary-layer thickness, which was computed through the procedure proposed by Vinuesa et al. [15], and the overbar denotes the boundary-layer thickness averaged over the region where β is observed to remain constant.

3 Results

The focus of this work is on obtaining a detailed characterization of the constant-β cases, which will ultimately allow to assess pressure-gradient effects with progressively more complex history effects, given by the particular $\beta(x)$ distribution. One of the two cases under consideration exhibits a constant value of $\beta \simeq 1$ in the range $910 < Re_\theta < 3360$, and the other one has a constant value of $\beta \simeq 2$ in the range $940 < Re_\theta < 4000$. In Fig. 1 we show a schematic representation of the constant $\beta = 1$ region, in comparison with the one obtained in the recent work by

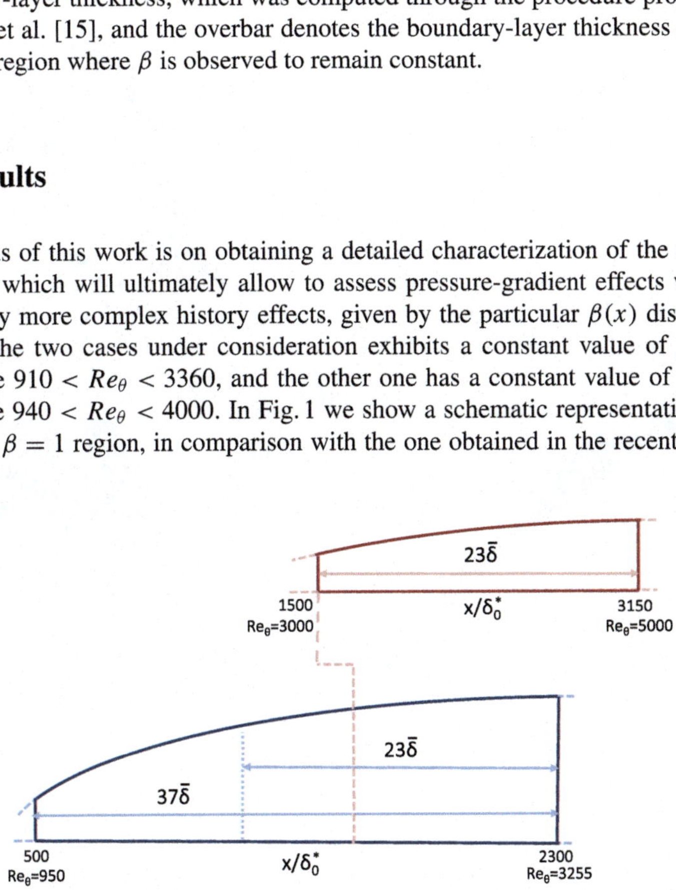

Fig. 1 —— Sketch of the $\beta = 1$ APG TBL case showing the area where a constant value of β was obtained, where δ_0^* is the displacement thickness of the laminar inflow boundary layer. —— Domain of interest with $\beta = 1$ extracted from the study by Kitsios et al. [9]. The extent of the constant $\beta = 1$ regions are shown in both cases normalized with the averaged boundary-layer thicknesses $\overline{\delta}$. The extent of the domain of interest from Kitsios et al.[9] is also represented in our case

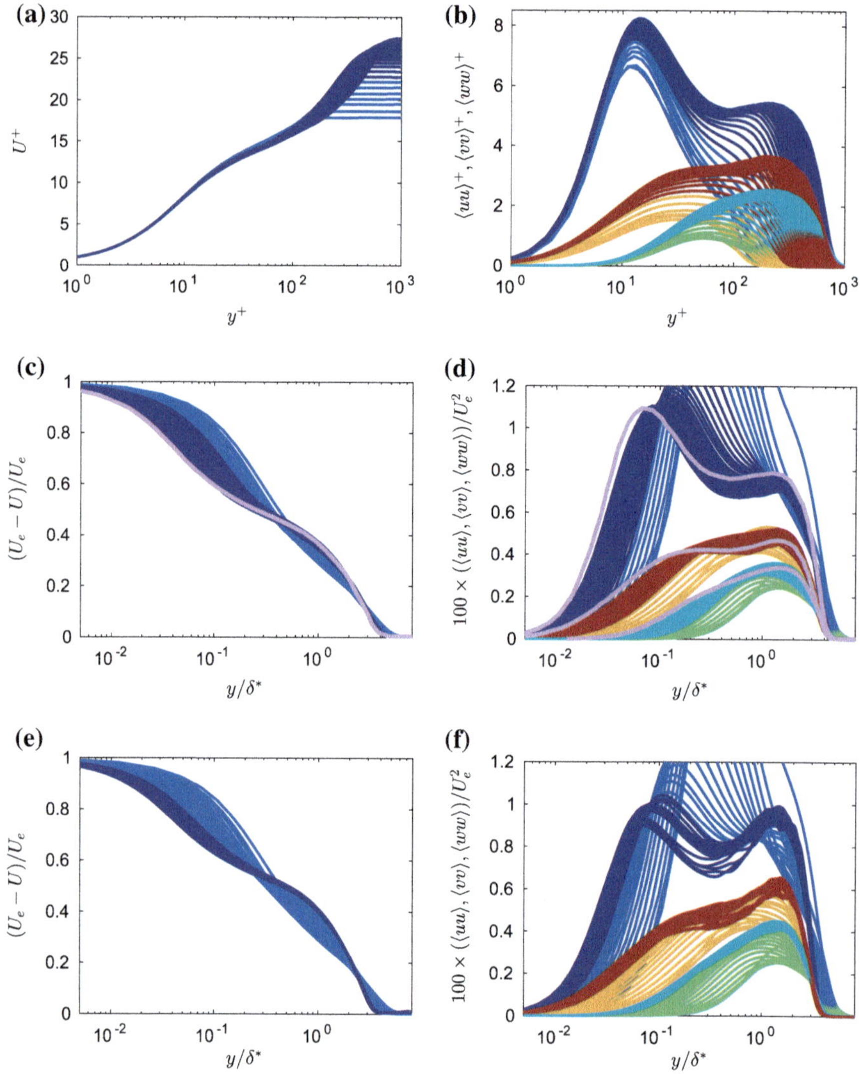

Fig. 2 Case $\beta = 1$: mean streamwise velocity profiles and selected Reynolds-stress tensor components (31 profiles in the range $100 < x < 2,300$) non-dimensionalised by **a**, **b** u_τ and $\ell^* = \nu/u_\tau$, and **c**, **d** U_e and δ^*. ——— Indicates profiles in the area of constant β for U and $\langle uu \rangle$, whereas ——— denotes profiles where β is not constant. ——— and ——— denote profiles of $\langle vv \rangle$ and $\langle ww \rangle$ within the constant β region, whereas ——— and ——— correspond to profiles without constant β. Average of collapsed profiles reported by Kitsios et al. [9] represented by ———. Case $\beta = 2$: **e** mean streamwise velocity profiles (23 positions in the range $100 < x < 2,300$), and **f** selected Reynolds-stress tensor components, with same scaling and color code as in **c** and **d**

Kitsios et al. [9], also for a constant $\beta = 1$ case. Note that although Kitsios et al. [9] explored higher Reynolds numbers than the ones considered here, the range over which β is constant is 1.6 times larger in the present simulation, and the relative Reynolds-number change is also larger.

Figure 2a and b show the inner-scaled streamwise mean and velocity fluctuation profiles corresponding to the $\beta = 1$ case, schematically discussed in Fig. 1. The profiles within the region of constant $\beta = 1$ are highlighted in the two panels through the darker colors. An alternative scaling for these quantities was considered by Kitsios et al. [9] based on the results from their simulation, in which the displacement thickness δ^* and the local edge velocity U_e were the characteristic length and velocity scales. They observed an apparent collapse of the mean flow and the fluctuations in their region of constant β, which as indicated in Fig. 1 corresponds to a streamwise distance of around 23 average boundary-layer thicknesses $\overline{\delta}$.

The scaling proposed by Kitsios et al. [9] is applied to our data in Fig. 2c and d, and as apparent no such collapse in any of the investigated quantities in our constant β region, which spans a longer streamwise distance of $37\overline{\delta}$ (and exhibits a larger increase in *Re*), is observable. One possible explanation for this discrepancy could be that the scaling considered by Kitsios et al. [9] does not lead to self-similarity, and since their constant β region is shorter than ours and their relative change in *Re* is narrower, their streamwise development would be insufficient to reveal this conclusion. The present data exhibits a clear *Re* trend, which is furthermore extended through the higher *Re* data by Kitsios et al. [9]. The failure of the scaling would indeed be in agreement with Townsend [1], since in principle the sink flow is the only flow that can be described from the wall to the freestream in terms of a single similarity variable in y. It is also interesting to note that recent particle image velocimetry (PIV) experiments performed in a similar configuration to that of Kitsios et al. [9], carried out by the same research group, also show *Re*-dependent profiles of the Reynolds-stress tensor components [16]. These aspects are further explored by analyzing the constant $\beta = 2$ case, over a streamwise distance of $28\overline{\delta}$. A higher Re_θ range is reached in this case, which is more comparable to the one analyzed by Kitsios et al. [9], albeit at a higher value of β. As seen from Fig. 2e and f, the scaling by Kitsios et al. [9] does not lead to self-similarity in this case either. Also here a clear *Re* trend is noticed, supporting the statements presented above, and also the validity of the classic two-layer similarity, at least for the β range under consideration.

4 Conclusions

In the present work we investigated the scaling proposed by Kitsios et al. [9], in which δ^* and U_e are considered as length and velocity scales. Our results show that this scaling does not lead to self-similar boundary layer profiles in the constant-β region. This conclusion is in agreement with Townsend [1], who showed that the sink flow is the only boundary layer exhibiting self-similarity. Although the current flat-plate simulations established long constant-β regions, stronger streamwise constant pressure gradients at higher Reynolds numbers should be investigated in order to characterise cases closer to wind-tunnel experiments and general applications.

Disclaimer: In parallel to the present paper, a largely extended and more detailed study based on the present work has been published by Bobke et al. [17]

References

1. A.A. Townsend, *The Structure of Turbulent Shear Flow* (Cambridge Univ. Press, Cambridge, UK, 1956)
2. G.L. Mellor, D.M. Gibson, Equilibrium turbulent boundary layers. J. Fluid Mech. **24**, 225–253 (1966)
3. P. Schlatter, R. Örlü, Q. Li, G. Brethouwer, J.H.M. Fransson, A.V. Johansson, P.H. Alfredsson, D.S. Henningson, Turbulent boundary layers up to $Re_\theta = 2500$ studied through simulation and experiment. Phys. Fluids **21**, 051702 (2009)
4. S.C.C. Bailey, M. Hultmark, J.P. Monty, P.H. Alfredsson, M.S. Chong, R.D. Duncan, J.H.M. Fransson, N. Hutchins, I. Marusic, B.J. Mckeon, H.M. Nagib, R. Örlü, A. Segalini, A.J. Smits, R. Vinuesa, Obtaining accurate mean velocity measurements in high Reynolds number turbulent boundary layers using Pitot tubes. J. Fluid Mech. **715**, 642–670 (2013)
5. P.E. Skåre, P.-Å. Krogstad, A turbulent equilibrium boundary layer near separation. J. Fluid Mech. **272**, 319–348 (1994)
6. M. Skote, D.S. Henningson, Direct numerical simulation of a separated turbulent boundary layer. J. Fluid Mech. **471**, 107–136 (2002)
7. P. Schlatter, R. Örlü, Turbulent boundary layers at moderate Reynolds numbers. Inflow length and tripping effects. J. Fluid Mech. **710**, 5–34 (2012)
8. R. Vinuesa, P.H. Rozier, P. Schlatter, H.M. Nagib, Experiments and computations of localized pressure gradients with different history effects. AIAA J. **55**, 368–384 (2014)
9. V. Kitsios, C. Atkinson, J.A. Sillero, G. Borrell, A.G. Gungor, J. Jiménez, J. Soria, Direct numerical simulation of a self-similar adverse pressure gradient turbulent boundary layer. Int. J. Heat Fluid Flow **61**, 129–136 (2016)
10. S.M. Hosseini, R. Vinuesa, P. Schlatter, A. Hanifi, D.S. Henningson, Direct numerical simulation of the flow around a wing section at moderate Reynolds number. Int. J. Heat Fluid Flow **61**, 117–128 (2016)
11. A. Bobke, R. Vinuesa, R. Örlü, P. Schlatter, Large-eddy simulations of adverse pressure gradient turbulent boundary layers. J. Phys. Conf. Ser. **708**, 012012 (2016)
12. M. Chevalier, P. Schlatter, A. Lundbladh, D.S. Henningson, SIMSON a pseudospectral solver for incompressible boundary layer. TRITA-MEK 2007:07, Technical Report, KTH Mechanics (2007)
13. P. Schlatter, S. Stolz, L. Kleiser, LES of transitional flows using the approximate deconvolution model. Int. J. Heat Fluid Flow **25**, 549–558 (2004)
14. G. Eitel-Amor, R. Örlü, P. Schlatter, Simulation and validation of a spatially evolving turbulent boundary layer up to $Re_\theta = 8300$. Int. J. Heat Fluid Flow **47**, 57–69 (2014)
15. R. Vinuesa, A. Bobke, R. Örlü, P. Schlatter, On determining characteristic length scales in pressure-gradient turbulent boundary layers. Phys. Fluids **28**, 055101 (2016)
16. C. Atkinson, A.J. Buchner, M. Eisfelder, V. Kitsios, J. Soria, Time-resolved PIV measurements of a self-similar adverse pressure gradient turbulent boundary layer. in, Proc. 18th Intl Symposium on the Application of Laser and Imaging Techniques to Fluid Mechanics, Lisbon, Portugal (2016)
17. A. Bobke, R. Vinuesa, R. Örlü, P. Schlatter, History effects and near equilibrium in adverse-pressure gradient turbulent boundary layers. J. Fluid Mech. **820**, 667–692 (2017)

Part III
Pipe Flows

Transitional and Turbulent Bent Pipes

Philipp Schlatter, Azad Noorani, Jacopo Canton, Lorenz Hufnagel, Ramis Örlü, Oana Marin and Elia Merzari

Abstract We review a number of aspects of the transitional and turbulent flow in bent pipes, obtained at KTH using the spectral-element code Nek5000. This flow, sometimes also called Dean flow, is characterised by the appearance of Dean vortices, which arise due to the action of the centrifugal force in the bend. We start with reviewing recent stability analysis in the toroidal flow, and conclude that for all curvatures $\delta > 0$ an exponential instability is present at a bulk Reynolds number of about 4000. Further increasing the Reynolds number lets the flow go through a region with potential sub straight and sublaminar drag. An analysis using proper orthogonal decomposition (POD) reveals that wave-like motions are still present in the otherwise turbulent flow. Upon further increasing Re, the in-plane Dean vortices lead to a modulation of turbulence depending on the azimuthal position. The flow is then dominated by low-frequency so-called swirl-switching motion. This motion is studied in both a periodic and spatially developing framework. Finally, the effect of Dean vortices on Lagrangian inertial particles is studied.

1 Introduction

The flow in bent pipes is an important natural extension of straight pipe flow, however, significantly less studies are devoted to bent pipes as compared to their straight counterparts. Due to the curvature, the azimuthal symmetry of the flow is broken, and centrifugal forces lead to the appearance of a secondary flow, i.e. an in-plane flow which manifests itself in the formation of two so-called Dean vortices. This secondary flow is skew-induced, and appears for both laminar and turbulent flow. The strength of these Dean vortices depends on the (bulk) Reynolds number Re_D, the curvature δ (usually defined as the ratio of pipe radius to the radius of the curvature), and also the streamwise extent of the bend. The latter parameter distinguishes the

P. Schlatter (✉) · A. Noorani · J. Canton · L. Hufnagel · R. Örlü
Linné FLOW Centre, KTH Mechanics, Stockholm, Sweden
e-mail: pschlatt@mech.kth.se

O. Marin · E. Merzari
MCS, Argonne National Laboratory, Lemont, USA

R. Örlü et al. (eds.), *Progress in Turbulence VII*, Springer Proceedings in Physics 196, DOI 10.1007/978-3-319-57934-4_12

flow in a torus (see e.g. [1]) from the one in spatially developing pipes, as for instance the 90° bend [2]. A recent review of both experiments and simulations in bent pipe configuration is provided in [3].

2 Numerical Setup

The fully resolved direct numerical simulations (DNS) are performed using Nek5000 [5], a high-order spectral-element code, in a similar way as discussed in e.g. [4, 6]. Special care has been taken to assure that the somewhat unusual spectral-element discretisation with its non-equidistant point distribution even in homogeneous (periodic) directions does not lead to any visible artifacts in the results. As an example, Fig. 1 shows an in-plane cut of the streamwise vorticity at moderate Reynolds number $Re_\tau = 1000$ [4]; in this case the Reynolds number is based on the friction velocity and pipe radius. The spectral-element mesh is Cartesian over the shown section, and involves curved elements. Nevertheless, no discontinuities at the elemental boundaries are present in the solution, which is completely continuous even for the vorticity which involves derivatives of the primary flow variables. This indicates that both the chosen resolution (fixed in inner units for the various Reynolds numbers considered) and the other details of the solver are suitable for accurate DNS of turbulent flows.

The diameter $D = 2R$ is used to define the bulk Reynolds number Re_D, which is set to $Re_D = 11,700$ for most cases reported in this article, which corresponds in the straight section to a friction Reynolds number of $Re_\tau \approx 360$. To assure that no non-

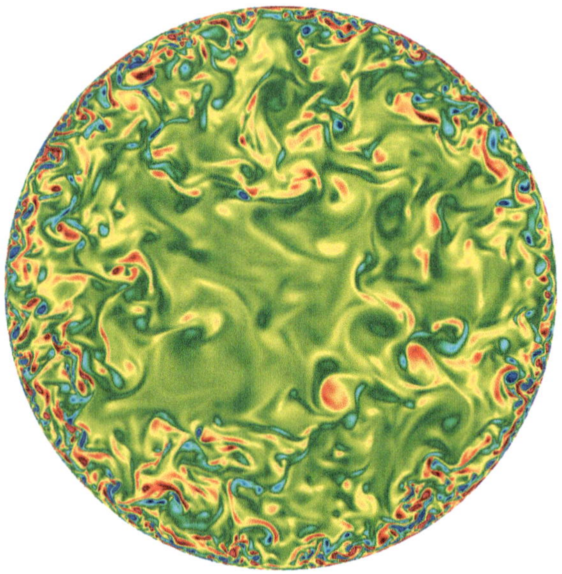

Fig. 1 Colour visualisation of the streamwise vorticity ω_z in a straight pipe at $Re_\tau = 1000$ [4]. Even though the mesh is non-uniform across the shown cut, no artifacts are visible even in the flow derivatives

physical effects are incurred throughout the simulation via the artificial periodicity, all shown torus simulations were performed in a pipe of total length of approximately $25R$. The spatially developing bend is even longer with a total length of $25D$.

The steady solutions and the stability analysis have been computed with PaStA, an in-house developed software code written in primitive variables and based on the finite element method (FEM, for details see [7]).

3 Stability and Transition

It has been known for a long time that the laminar flow in straight pipes does not exhibit a linear instability at any "relevant" Reynolds number (even though there is no formal proof as opposed to Couette flow). However, a similar analysis has not been performed for pipes with curvature, even though some recent experiments and simulations suggest a wave-like instability. Therefore, we started studying the laminar flow in bent pipes [8], and confirmed that indeed the curvature and the Reynolds number need to be considered independent parameters, and cannot be collapsed into a single Dean number. In [9] we provide a complete linear stability analysis, and found that for any pipe with curvature larger than zero exhibits a linear instability at $Re \approx 4000$; the corresponding stability diagram is shown in Fig. 2, and is composed of a multitude of modes grouped in different families. Note that for lower curvatures also subcritical transition, as observed in straight pipes, has been observed, but is not shown in the diagram.

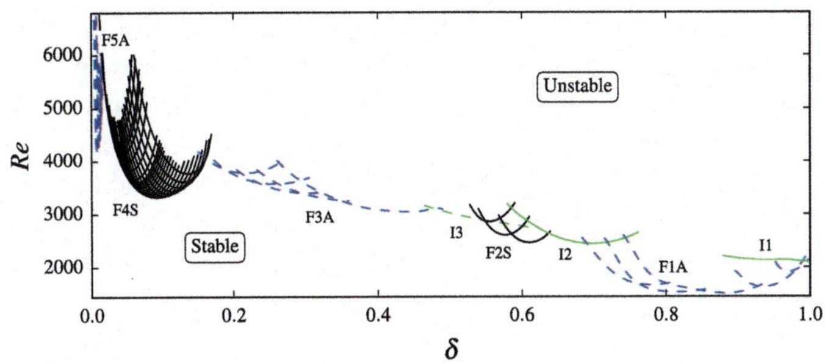

Fig. 2 Neutral curve in the $\delta - Re$ plane for $\delta \in [0.002, 1]$, see [9]. Each line corresponds to the neutral curve of one mode. The neutral curve for the flow is formed by the envelope of the lines. Five families *(black and blue)* and three isolated modes *(green)* are marked by labels. Symmetric modes are indicated with *continuous lines* while antisymmetric modes are represented with *dashed lines*. Note that the curves are not interpolated, i.e. they are segments connecting computed solutions with $\Delta\delta = \mathcal{O}(10^{-3})$. The uncertainty on the Reynolds number is $\pm 10^{-4}$

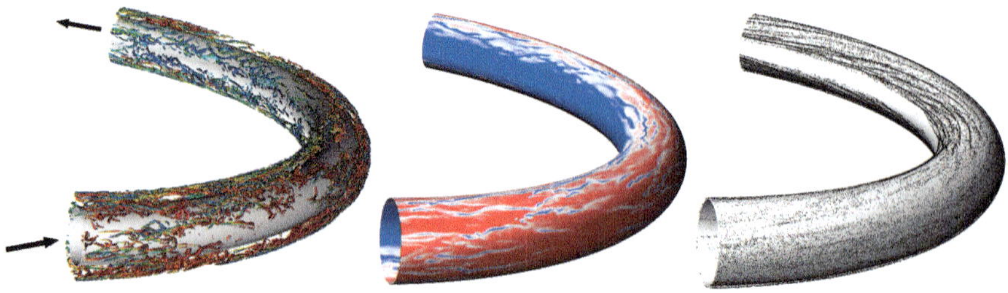

Fig. 3 Turbulence in a bent pipe: The pictures show from *left to right*: Vortical structures in the near-wall region, coloured with the streamwise velocity [6]; the wall-shear stress indicating partial laminarisation at the inner bend [12]; particle distribution for inertial particles [11]. For all simulations $Re_\tau = 360$, curvature $\delta = 0.1$

4 Turbulent Flow in Bent Pipes

After the flow has undergone transition, a turbulent flow is established [6] which is modulated by the in-plane Dean flow. This means that the turbulence at the outer side of the bend is enhanced, and partial laminarisation is observed at the inner side of the bend. A typical snapshot of such a turbulent flow at intermediate curvature is shown in Fig. 3, showing the azimuthal dependence of turbulence. Motivated by this inhomogeneity, we also studied inertial Lagrangian particles in bent pipes, and concluded that a bend may have a crucial impact on the spatial distribution of particles, see [10, 11]. For specific conditions there are regions, located in the centre of the Dean vortices, that are never visited by any particle. This knowledge may be important when designing probes that measure e.g. concentration.

It is intuitively clear that the drag induced by the flow in a bent pipe is generally larger than in the straight counterpart at the same mass flux. However, there exists a regime at comparably low Reynolds number and curvature, where this is not the case: both sub-straight drag and sub-laminar drag could be established using our numerical simulations [12]. It turns out that in these configurations the bend induces comparably strong wave-like motion in the flow, which transports the energetic near-wall flow towards the centre of the Dean vortices, thereby reducing the wall gradient.

5 Swirl Switching

The flow in bent pipes at high Reynolds number and sufficiently large curvatures has been known to exhibit low-frequency oscillations, which may contribute to fatigue of the structure [13]. This so-called swirl switching is the periodic dominance of one

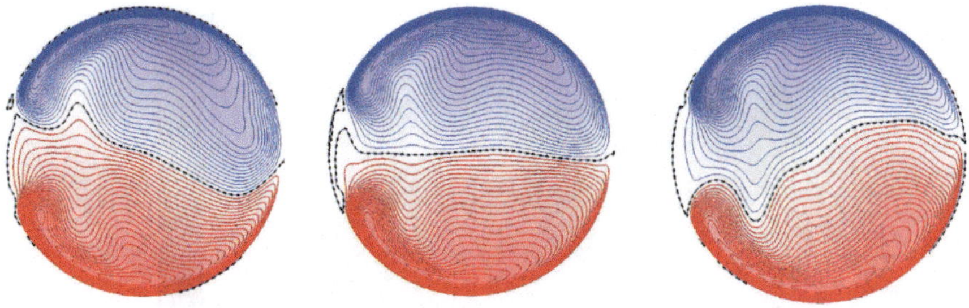

Fig. 4 Illustration of the swirl switching as an alternate dominance of one Dean vortex over the other; shown are contours of the in-plane stream function for three time instants. The *inner side* of the bend is on the *left-hand side* [1]

Dean cell over the other, and has been the subject of a number of recent (experimental) papers. Using our simulation setup we wanted to study this so-called swirl-switching phenomenon as well. We consider two different geometries, i.e. a toroidal pipe [1] and a spatially developing pipe [2].

For the torus simulations, an analysis using proper orthogonal decomposition (POD) revealed that the low-frequency oscillations characteristic of swirl switching could be detected in the flow [1]; Fig. 4 shows a low-order reconstruction with the 50 most energetic modes. This result indicates that in order for swirl switching to happen, no upstream straight section of the pipe is necessary, and only the bent part is sufficient. However, in experiments only spatially developing bends are considered, therefore we aimed at setting up a similar case in order to perform POD analysis on a spatial bend as well, see Fig. 5. In order to generate a turbulent inflow for the spatially developing bend, the synthetic eddy method [14] has been adapted to the current setup. This choice was particularly important in order to avoid spurious frequencies in the flow that would arise when using recycling conditions. Our experience with the current inflow method is very good, and the flow can be considered a canonical turbulent pipe flow already after 5 diameters downstream of the inflow. The POD analysis performed on the DNS data is classical; the only noteworthy aspect is that the symmetry of the flow (mirror symmetry) is exploited in our decomposition. In order to be able to compare to experiments, we have performed both 2D-POD (in cross-flow planes) and 3D-POD. It turns out that we could exactly reproduce all modes found in experiments using the 2D-POD technique. The fully three-dimensional POD modes reveal that the swirl switching is essentially one travelling mode, originating in the bend without connection to the inflowing turbulent flow. This finding highlights the importance of using the full three-dimensional velocity snapshots in order to extract correct modes, and thus an accurate low-order description of the phenomenon.

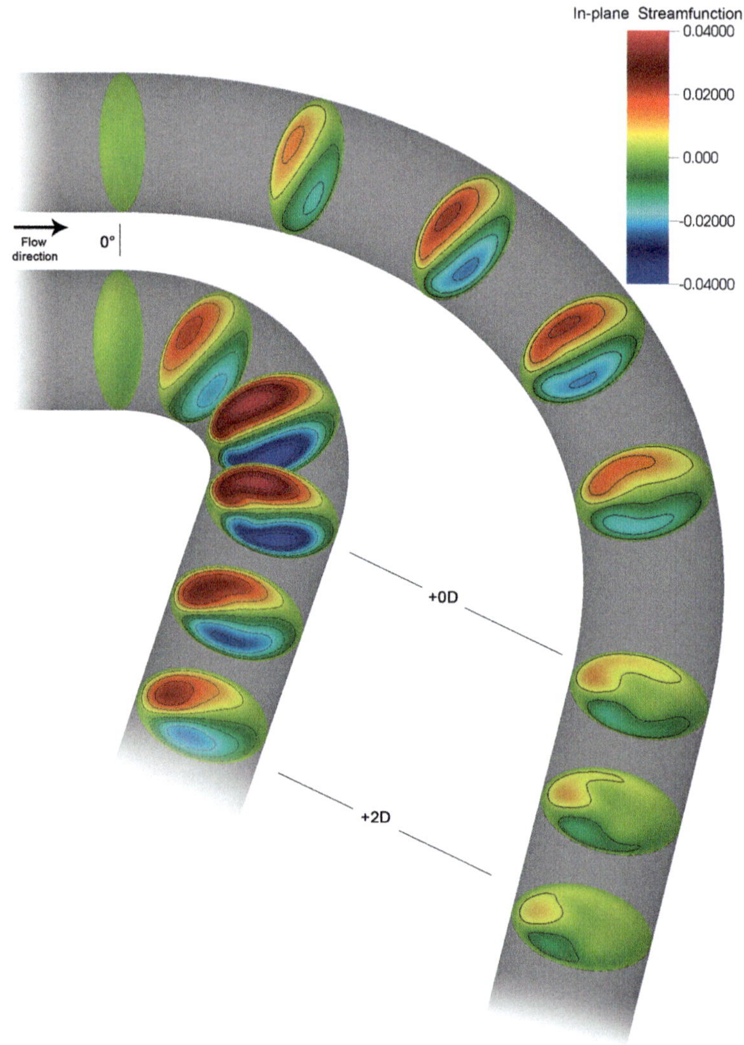

Fig. 5 Setup of the spatially varying bent pipe for studying the swirl switching: The flow enters on the top left, and is curved by 90° with curvature $\delta = 0.1$ and 0.3. The colours indicate the amplitude of the inplane streamfunction. The appearance of Dean vortices in the bend, and their subsequent decay can clearly be appreciated

Acknowledgements Financial support by the Swedish Research Council (VR) and the Knut and Alice Wallenberg Foundation (KAW) is gratefully acknowledged. Most of the simulations were performed on resources provided by the Swedish National Infrastructure for Computing (SNIC) at the National Supercomputer Centre (NSC) and PDC (Stockholm). Additional simulations were performed within the DECI-project PIPETURB.

References

1. A. Noorani, P. Schlatter, Swirl-switching phenomenon in turbulent flow through toroidal pipes. Int. J. Heat Fluid Flow **61**, 108–116 (2016)
2. L. Hufnagel, On the swirl switching in developing bent pipe flow with direct numerical simulation, Master's Thesis, KTH Mechanics, 2016
3. A. Kalpakli Vester, R. Örlü, P.H. Alfredsson, Turbulent flows in curved pipes: recent advances in experiments and simulations. Appl. Mech. Rev. **68**(5), 050802 (2016)
4. G.K. El Khoury, P. Schlatter, A. Noorani, P.F. Fischer, G. Brethouwer, A.V. Johansson, Direct numerical simulation of turbulent pipe flow at moderately high Reynolds number. Flow Turbul. Combust. **91**, 475–495 (2013)
5. P.F. Fischer, J.W. Lottes, S.G. Kerkemeier, Nek5000 Web page, 2008, http://nek5000.mcs.anl.gov
6. A. Noorani, G.K. El Khoury, P. Schlatter, Evolution of turbulence characteristics from straight to curved pipes. Int. J. Heat Fluid Flow **41**, 16–26 (2013)
7. J. Canton, Global linear stability of axisymmetric coaxial jets, Master's Thesis, Politecnico di Milano, 2013
8. J. Canton, R. Örlü, P. Schlatter, Characterisation of the steady, laminar incompressible flow in toroidal pipes covering the entire curvature range, submitted (2017)
9. J. Canton, P. Schlatter, R. Örlü, Modal instability of the flow in a toroidal pipe. J. Fluid Mech. **792**, 894–909 (2016)
10. A. Noorani, G. Sardina, L. Brandt, P. Schlatter, Particle velocity and acceleration in turbulent bent pipe flows. Flow Turbul. Combust. **95**(2), 539–559 (2015)
11. A. Noorani, G. Sardina, L. Brandt, P. Schlatter, Particle transport in turbulent curved pipe flow. J. Fluid Mech. **793**, 248–279 (2016)
12. A. Noorani, P. Schlatter, Evidence of sublaminar drag naturally occurring in a curved pipe. Phys. Fluids **27**, 035105 (2015)
13. C. Brücker, A time-recording DPIV-study of the swirl switching effect in a 90° bend flow. *8th International Symposium on Flow Visualization*, Sorrento, Italy p. 171 (1998)
14. N. Jarrin, S. Benhamadouche, D. Laurence, R. Prosser, A synthetic eddy-method for generating inflow conditions for large-eddy simulations. Int. J. Heat Fluid Flow **27**(4), 585–593 (2006)

Turbulent Pipe Flow Near-Wall Statistics

Tommaso Fiorini, Gabriele Bellani, Ramis Örlü, Antonio Segalini, P. Henrik Alfredsson and Alessandro Talamelli

Abstract Results from the first experimental campaign in the Long Pipe facility of the CICLoPE laboratory are reported. Single hot-wire profile measurements are presented, taken from the wall up to one third of the pipe radius, with the friction Reynolds number Re_τ ranging from 6.5×10^3 up to 3.8×10^4. Measurements of the pressure drop along the pipe are presented together with an estimation of its uncertainty. Mean and variance of the streamwise velocity fluctuations are examined and compared with the findings from other facilities. The amplitude of the inner-scaled near-wall peak of the variance, after being corrected for spatial resolution effects, shows an increasing trend with Reynolds number, in accordance with low Reynolds number experiments and simulations.

1 Introduction

Turbulent pipe flow is one of the canonical wall flows, and it has been the object of a multitude of studies over the years. Direct numerical simulations (DNS) and experiments have progressively led to an improved understanding of wall turbulence, but despite being intensively studied, many issues remain unresolved, such as the Re-scaling of the variance profile (see [1–3]). Laboratory experiments continue to play an essential role in the understanding of the physics of wall-bounded turbulence, since DNS are still limited to low Re. The CICLoPE Long Pipe facility [4] aims at shedding light on some of the open questions of wall turbulence. The facility is unique in its kind, due to its very large dimensions, high Re can for the first time be reached while still maintaining a sufficient spatial resolution to resolve all the scales of turbulent motion with conventional hot-wire sensors. Spatial filtering is a great challenge in experiments [5], and has been masking the true nature of turbulence at high Re. Beside the resolution, the facility's high degree of manufacturing tolerances and

T. Fiorini · G. Bellani · A. Talamelli (✉)
DIN, Universita' di Bologna, 47100 Forli', Italy
e-mail: alessandro.talamelli@unibo.it

R. Örlü · A. Segalini · P.H. Alfredsson
Linné FLOW Centre, KTH Mechanics, 10044 Stockholm, Sweden

R. Örlü et al. (eds.), *Progress in Turbulence VII*, Springer Proceedings in Physics 196, DOI 10.1007/978-3-319-57934-4_13

flow stability provide the possibility to carry out accurate and resolved experiments at high Re, in a way that has so far not been possible in any other wall-turbulence facility.

2 Experimental Setup

The Long Pipe facility in the CICLoPE laboratory is a closed-loop wind tunnel, where at the end of the test section, a 111 m long carbon-fiber pipe, a fully developed turbulent flow condition is reached. The pipe has an inner diameter of 901 ± 0.1 mm, resulting in a length-to-diameter ratio of $L/D \approx 123$. The dimension of the facility is the result of the sizing process detailed in [4]. The wind tunnel is also equipped with flow conditioning elements to ensure a good and stable flow quality; these include a heat exchanger, a honeycomb, 5 screens, a settling chamber, and a convergent with contraction ratio of 4; each of the six corners of the loop is also equipped with turning vanes. The wind tunnel is driven by two-stage axial fans for a total power of 480 kW, however for the present measurements only one fan was used and the other was free running. For technical details about the final design of the facility and its elements, the reader is referred to [6]. In Fig. 1 an overview of the facility with its principal elements is shown. Hot-wire anemometry measurements are performed close to the end of the test section at $L/D = 122$, as a part of the same experimental campaign described in [4], but here a new hot-wire data-set with a shorter sensor is shown. The data presented here are acquired with a custom-made boundary-layer type probe, with a 1.2 μm diameter platinum wire soldered on stainless steel prongs; the wire length is 0.25 mm in order to keep a wire aspect ratio of $l/d \approx 200$. The hot-wire is operated in constant-temperature mode via a Dantec Streamline system. The sampling frequency is set to 60 kHz with an analog low-pass filter at 30 kHz for all cases. Velocity calibration was performed ex situ in a DANTEC Streamline 90H02 external calibrator jet. The hot-wire probe is mounted on a traversing system that consists of a hollow carbon-fiber airfoil that slides through the pipe wall (see Fig. 2) and spans from the wall up to $y/R \approx 0.3$; where y is the wall-normal distance and R is the pipe radius. The probe is traversed via a stepper motor with a 5 μm resolution step, while the relative position is obtained using a Renishaw Tonic T100x

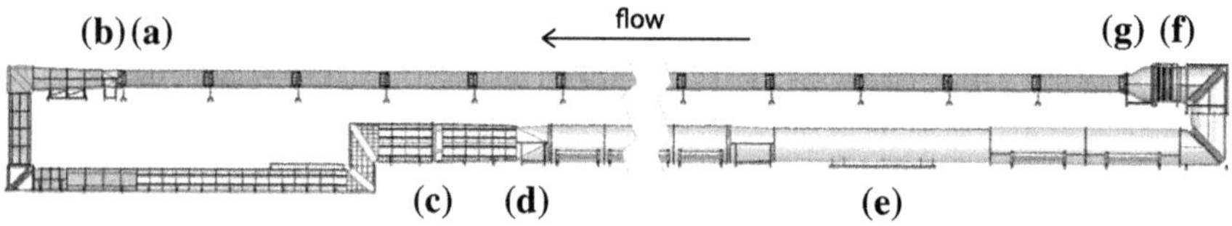

Fig. 1 Overview of the Long Pipe flow loop. **a** Measuring station. **b** Round to rectangular shape converter. **c** Heat exchanger. **d** Rectangular to round shape converter. **e** Axial fans. **f** Flow conditioning unit (honeycomb, screens). **g** Convergent with contraction ratio 4

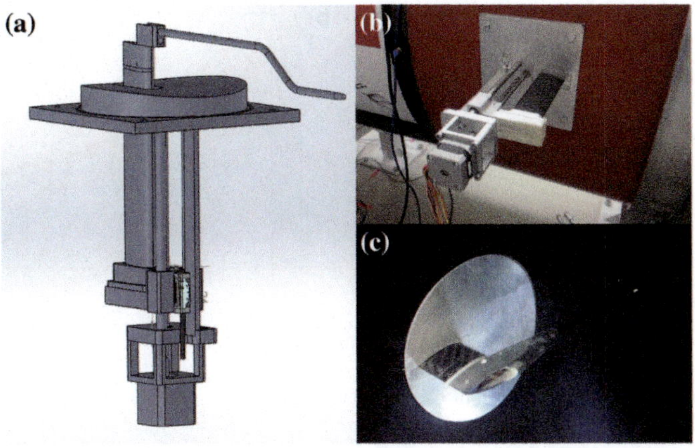

Fig. 2 The traversing device used for the hot-wire measurements, described in Sect. 2. **a** CAD drawing of the device. **b** Traversing as seen from outside the pipe, including the stepper motor and optical encoder. **c** Traversing as seen from inside the pipe, showing the probe holder and sting assembly

optical linear encoder with a 0.5 μm resolution. The mean centreline velocity is measured with a Prandtl tube connected to a MKS Baratron 120AD differential pressure transducer with a 1333 Pa range. The ambient pressure and temperature inside the test chamber are acquired with a MKS Baratron 120A absolute pressure transducer and a PT100 platinum thermoresistor, respectively. The pressure along the pipe is acquired through 16 static pressure taps, with a hole diameter of 1 mm, connected to a digital pressure scanner Initium with 2500 Pa range.

3 Pressure Drop Measurement

The pressure taps used to determine the pressure drop inside the pipe are located from the hot-wire measuring station up to 70 m upstream, with a 5 m spacing between each of them. To obtain the pressure gradient dp/dx and therefore the wall friction τ_w, a least-square fitting is performed. In order to determine the region and number of pressure taps to use, different linear fits of the data have been made starting from the test section and moving upstream using an increasing number of points for a wide range of friction Reynolds number Re_τ, where $Re_\tau = u_\tau R/\nu$, with u_τ denoting the friction velocity and ν the kinematic viscosity of the fluid. To evaluate the quality of those fits, their uncertainty can be computed, as shown in Fig. 3a. The more measurement points are used, the less the fit is influenced by bias errors introduced by single pressure taps; on the other hand, a local measure at the hot-wire station is desired, and an overall lower uncertainty of the fit does not necessarily mean that

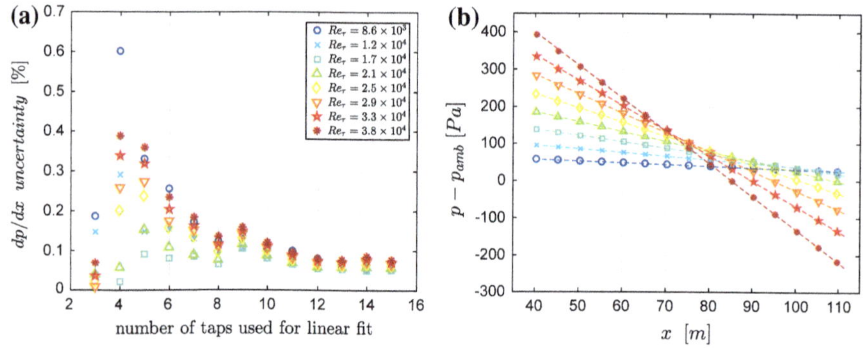

Fig. 3 **a** Uncertainty on the linear fit used to determine dp/dx, as a function of the number of pressure taps used for the fit. **b** Difference between the static pressure measured along the pipe and the ambient pressure, where x indicates the axial distance from the start of the pipe. *Dashed lines* indicate the linear fits obtained by using the last 6 points

the dp/dx at the hot-wire station is measured more accurately. Only relatively small benefits are visible by adding more data points beyond the 6th–7th tap. It was thus decided to compute the pressure gradient using the last 6 pressure taps corresponding to the last 30 m of the pipe, to avoid any residual flow development effect. Static pressure data points and corresponding linear fits calculated in this way are shown in Fig. 3b for a range of Reynolds number.

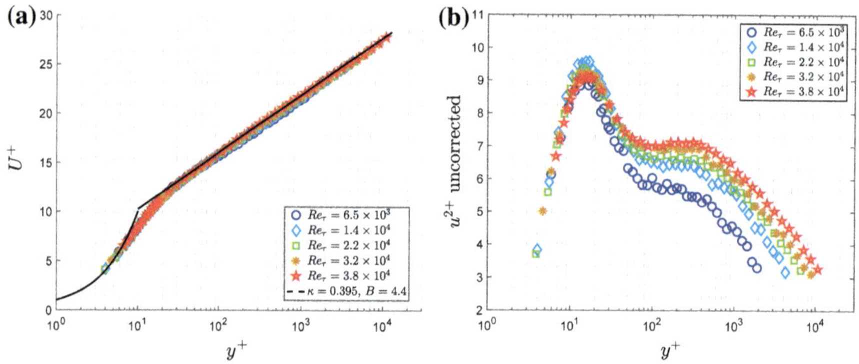

Fig. 4 **a** Inner-scaled mean velocity profile. *Solid black line* shows a logarithmic law with coefficients given in the legend. **b** Inner-scaled streamwise velocity variance without any correction applied

4 Near-Wall Statistics

Single-wire velocity profiles were acquired with the traversing system described in Sect. 2, the pressure drop was acquired and averaged for the duration of the profile acquisition. A calibration was performed before and after each profile to check for sensor drift. The mean velocity normalized with the friction velocity, U^+, is shown in Fig. 4a. The absolute wall position was retrieved by fitting the data points for $U^+ < 10$ using an analytical expression for the law of the wall [7]. The friction velocity was obtained directly from the pressure drop as outlined in Sect. 3. Data collapse is satisfying with a slight deviation on the dataset at the lowest *Re*, i.e. $Re_\tau = 6.5 \times 10^3$, which might be related to insufficient calibration data points for the lowest velocity range, or to the increased uncertainty in the pressure drop determination. The normalized mean velocity profiles are shown together with the linear relationship of the viscous sublayer and a reference logarithmic law with coefficients $\kappa = 0.395$ and $B = 4.40$. It should be noted that an accurate analysis of the mean velocity logarithmic region, its limits and the value of the coefficients has yet to be performed, and the value of the coefficients given here are just for reference. The coefficients of the log law also appear to differ slightly from what is measured in [8], where different hot-wire data-sets from the same experimental campaign are presented. The measured streamwise velocity variance normalized with the friction velocity, u^{2+}, is shown in Fig. 4b; data was then corrected using the semi-empirical expression reported in [9] and the corrected variance is shown in Fig. 5a. The corrected u^{2+}, shows a clear trend in the magnitude of the near-wall peak with increasing Reynolds number, particularly visible at lower Re_τ, as such confirming the findings from channel and boundary layers [5, 10], while differing from observations from the Princeton University/ONR Superpipe facility [11, 12]. The *Re*-trend of

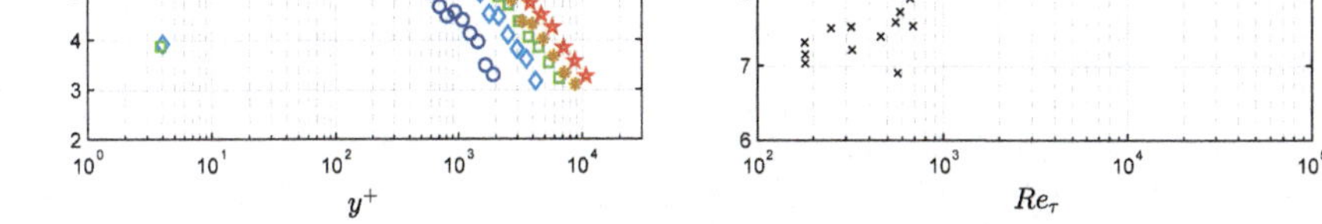

Fig. 5 **a** Inner-scaled streamwise velocity variance after being corrected with the scheme proposed in [9] **b** Amplitude of the variance near-wall peak $u^{2+}|_m$, as a function of the friction Reynolds number. $\times$ symbols represent data reported in [14], Δ symbols are the data from present measurements whereas the *coloured* symbols show the same data corrected for spatial averaging using [9]

the amplitude of the near-wall peak is given in Fig. 5b together with some results measured in the KTH pipe facility [13, 14]. As far as the 'second peak' observed in other pipe flow experiments [11, 12, 15, 16] and predicted in [17] is concerned, the present results cannot confirm its presence, as no clear peak is visible up until the highest Reynolds number investigated here ($Re_\tau = 3.8 \times 10^4$). Nevertheless, it should be noted that in the Superpipe measurements, at these Reynolds number, such a peak has only started to appear and is still quite 'subtle' in appearance.

Acknowledgements Financially supported through the European High-Performance Infrastructures in Turbulence (EuHIT) within the Reynolds stress tensor scaling in turbulent pipe flow (Re-Scale) project.

References

1. I. Marusic, B.J. McKeon, P.A. Monkewitz, H.M. Nagib, A.J. Smits, K.R. Sreenivasan, Wall-bounded turbulent flows at high Reynolds number: recent advance and key issues. Phys. Fluids **22**, 065103 (2010)
2. J. Kim, Progress in pipe and channel flow turbulence 1961–2011. J. Turbul. **13**, N45 (2012)
3. A.J. Smits, B.J. McKeon, I. Marusic, High-Reynolds number wall turbulence. Annu. Rev. Fluid Mech. **43**, 353–375 (2011)
4. A. Talamelli, F. Persiani, J.H.M. Fransson, P.H. Alfredsson, A.V. Johansson, M. Nagib, H. Rüedi, J.-D. Sreenivasan, P.A. Monkewitz, CICLoPE—a response to the need for high Reynolds number experiments. Fluid Dyn. Res. **41**, 021407 (2009)
5. N. Hutchins, T.B. Nickels, I. Marusic, M.S. Chong, Hot-wire spatial resolution issues in wall-bounded turbulence. J. Fluid Mech. **635**, 103–136 (2009)
6. G. Bellani, A. Talamelli, The final design of the long pipe in CICLOPE, in *Springer Proceedings in Physics, Progress in Turbulence VI*, Springer, pp. 205–209 (2016)
7. K.A. Chauhan, P.A. Monkewitz, H.M. Nagib, Criteria for assessing experiments in zero pressure gradient boundary layers. Fluid Dyn. Res. **41**, 021404 (2009)
8. R. Örlü, T. Fiorini, G. Bellani, A. Segalini, P.H. Alfredsson, A. Talamelli, Reynolds stress scaling in pipe flow turbulence—first results from CICLoPE. Phil. Trans. R. Soc. A **375**, 20160187 (2017)
9. A.J. Smits, J. Monty, M. Hultmark, S.C.C. Bailey, N. Hutchins, I. Marusic, Spatial resolution correction for wall-bounded turbulence measurements. J. Fluid Mech. **676**, 41–53 (2011)
10. P.H. Alfredsson, R. Örlü, A. Segalini, A new formulation for the streamwise turbulence intensity distribution in wall-bounded turbulent flows. Eur. J. Mech. B/Fluids **36**, 167–175 (2012)
11. M. Hultmark, S.C.C. Bailey, A.J. Smits, Scaling of near-wall turbulence in pipe flow. J. Fluid Mech. **649**, 103–113 (2010)
12. M. Hultmark, M. Vallikivi, S.C.C. Bailey, A.J. Smits, Turbulent pipe flow at extreme Reynolds numbers. Phys. Rev. Lett. **779**, 371–389 (2012)
13. M. Ferro, Experimental study on turbulent pipe flow. M.Sc. Thesis, KTH Mechanics, Royal Institute of Technology, Stockholm (2012)
14. R. Örlü, P.H. Alfredsson, Comment on the scaling of the near-wall streamwise variance peak in turbulent pipe flows. Exp. Fluids **54**, 1431 (2013)
15. J.F. Morrison, B.J. McKeon, W. Jiang, A.J. Smits, Scaling of the streamwise velocity component in turbulent pipe flow. J. Fluid Mech. **508**, 99–131 (2004)
16. M. Vallikivi, M. Hultmark, S.C.C. Bailey, A.J. Smits, Turbulence measurements in pipe flow using a nano-scale thermal anemometry probe. Exp. Fluids **51**, 1521–1527 (2011)
17. P.H. Alfredsson, A. Segalini, R. Örlü, A new scaling for the streamwise turbulence intensity in wall-bounded turbulent flows and what it tells us about the "outer" peak. Phys. Fluids **23**, 041702 (2011)

High Reynolds Number Experimental Facilities for Turbulent Pipe Flow at NMIJ

Noriyuki Furuichi, Yoshiya Terao and Yoshiyuki Tsuji

Abstract In this paper, we report on high Reynolds number and highly accurate experimental facilities for turbulent pipe flow established by the National Metrology Institute of Japan (NMIJ). One of the facilities, called the High Reynolds number actual flow facility (Hi-Reff), is capable of handling a maximum bulk Reynolds number of $Re_D = 2.0 \times 10^7$. The most remarkable feature of this facility is its highly accurate flow rate measurements. The expanded uncertainty of the volumetric flow rate is estimated as 0.040–0.10%. Such a low flow rate measurement uncertainty contributes to extremely accurate estimations of inner-scale variables such as friction velocity. This paper presents the details and advantages provided by this NMIJ facility in relation to turbulent pipe experiments.

1 Introduction

High Reynolds number experiments for wall-bounded flows have contributed to the clarification of numerous turbulent characteristics, especially the universality of the profiles of mean and turbulent statistics. For pipe flows, Nikradse [1] performed experiments in the high Reynolds number region at 1930s, and recent experiments by Superpipe at Princeton University [2] and the Center for International Collaboration on Long Pipe Experiments (CICLoPE) at Bologna [3] are well known. In general, high Reynolds number experiments involve numerous issues including straight pipe length, surface roughness, spatial resolution, and measurement accuracy. To date, however, there have been neither experiments nor experimental facilities capable of handling all issues related to high Reynolds numbers.

National Metrology Institute of Japan (NMIJ) established the high Reynolds number flow facility in 2009 as a part of national standard for water flow rates in Japan [4].

N. Furuichi (✉) · Y. Terao
National Institute of Advanced Industrial Science and Technology (AIST), National Metrology Institute of Japan (NMIJ), 1-1-1 Umezono, Tsukuba, Japan
e-mail: furuichi.noriyuki@aist.go.jp

Y. Tsuji
Nagoya University, Furo-cho, Chikusa-ku, Nagoya, Japan

R. Örlü et al. (eds.), *Progress in Turbulence VII*, Springer Proceedings in Physics 196, DOI 10.1007/978-3-319-57934-4_14

This facility has achieved a maximum bulk Reynolds number of $Re_D = 2.0 \times 10^7$ with an expanded uncertainty level of 0.040–0.10% ($k = 2$) for the volumetric flow rate. In terms of what this very low flow rate uncertainty measurement brings to highly accurate inner-scale variables such as the friction velocity, the primary advantage of this facility is its ability to handle sophisticated turbulent pipe flow experiments. In fact, using this facility, the authors have already performed highly accurate friction factor and velocity profile experiments, the results of which have been already published in our previous paper [5]. However, in that paper, the specifications of the experimental facility used for our turbulent flow experiments were not elucidated. Accordingly, in this paper, the details and advantages provided by the above mentioned NMIJ facility for turbulent experiments will be presented.

2 Experimental Facility at NMIJ

An overview of the facility at NMIJ is shown in Fig. 1 and pictures of the facility are shown in Fig. 2. This facility is located in the Tsukuba North Site of the National Institute of Advanced Industrial Science and Technology (AIST) as shown in Fig 2a. The facility is approximately 200 m long and 50 m wide. To facilitate experiments over a wide range of standard flow rates, this facility incorporates several flow loops and weighing tanks with capacities from 10 kg to 50 t. For turbulent pipe flow experiments, two facilities are available; one is Hi-Reff, and the other, which is equipped with both overflow head and weighing tanks, is called the water flow facility. The working fluid of these facilities is water. Both facilities are equipped with important specifications and capabilities that facilitate turbulent pipe flow experiments such as a long straight pipe, high flow rate stability, and high temperature and water pressure stability. The most remarkable facet of both facilities, which provides an advantage over other experimental facilities, is their highly accurate flow rate measurements. This is achieved via the same static gravimetric method that is generally used for the flow meter calibrations.

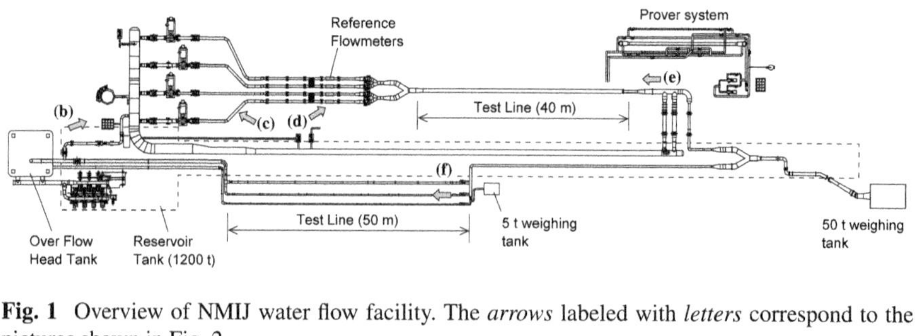

Fig. 1 Overview of NMIJ water flow facility. The *arrows* labeled with *letters* correspond to the pictures shown in Fig. 2

Fig. 2 High Reynolds number experimental facilities: **a** overview of entire facility. The facilities are located in the center of the figure. **b** Rear view of Hi-Reff. **c** Hi-Reff feed pumps and pressure tank. **d** Hi-Reff Reference flow meters section. **e** Hi-Reff test line. **f** Water flow facility test lines

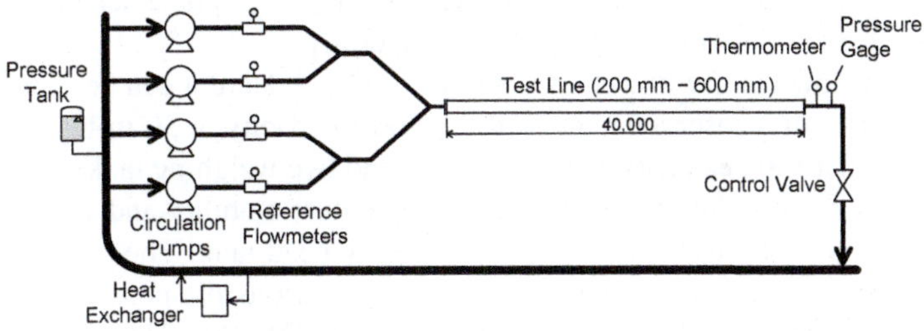

Fig. 3 Flowchart of Hi-Reff at NMIJ

The Hi-Reff flowchart is shown in Fig. 3. This facility is located outdoors and almost all of the piping is covered with insulation, as shown in Fig. 2b. Flows are fed to the test lines by the circulation pumps, which are shown in Fig. 2c. Since the maximum flow rate of each pump is 0.83 m^3/s, simultaneous operation of the four circulation pumps can generate water flows up to 3.33 m^3/s in the test line. The straight length of the test line is approximately 40 m (100*D* for DN400, 67*D* for DN600) and the available pipe diameters range from 200 to 600 mm. Figure 2e shows a photo of the Hi-Reff test line taken from the downstream side. The bulk velocity in the pipe depends on the pipe diameter and can be as high as 18 m/s. As stated in the experimental pipe example discussed in our previous paper, the pipe roughness is 0.25 μm and the roundness is less than 25 μm [5].

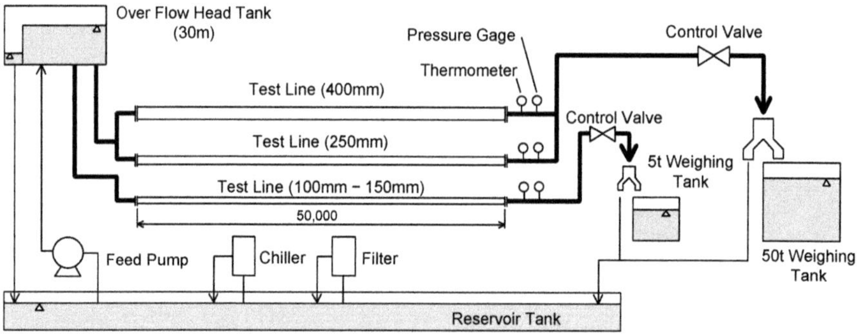

Fig. 4 NMIJ water flow facility flowchart

By using the boiler and the heat exchanger, the temperature is controllable from 20 to 75 °C at a stability within ±0.1 °C, even though almost all of this portion of the facility is located outside and thus exposed to the environment. In addition, line pressure is controlled from 0.2 to 0.4 MPa by the pressure tank, and at a stability level better than 0.001 MPa. The maximum Reynolds number based on the bulk velocity is $Re_D = 2.0 \times 10^7$, while the friction velocity is $Re_\tau = 3.0 \times 10^5$. The flow rate is given by the reference flow meters (Fig. 2d), which are calibrated by the static gravimetric method. The expanded uncertainty of the volumetric flow rate is estimated to be 0.10% with the coverage factor $k = 2$. For a schematic of the uncertainty estimation, refer to the previous paper [4].

The water flow facility flowchart is shown in Fig. 4. Here it can be seen that the facility consists of a reservoir tank with a 1200 t capacity, a 30 m-high overflow head tank with a 150 t capacity, three test lines, and two weighing tanks. Chiller and filter systems are installed to regulate the temperature stability and water density. The flow is fed to the test lines from the overflow head tank. The maximum flow rate is 0.83 m^3/s, and test piping is available in various diameters from 100 to 400 mm. The straight pipe length is 50 m (125D for DN400). The pipe roughness is less than 0.1 μm and roundness is less than 10 μm [5]. Water temperature is maintained at approximately 20 °C and the line pressure is kept less than 0.24 MPa, depending on flow rate. The maximum Reynolds number is $Re_D = 2.6 \times 10^6$ and $Re_\tau = 4.7 \times 10^4$. At the downstream end of the test lines, the weighing tanks are installed. The volumetric flow rate expanded uncertainties with $k = 2$ are estimated to be 0.060 and 0.042% for the 50 and 5 t weighing tank systems, respectively.

The experimental range of the NMIJ facilities, presented by the viscous length and the Reynolds number, is shown in Fig. 5. Since the pipe diameter of the NMIJ facilities is changeable, as mentioned earlier, the shaded and hatched areas show the available range. This figure shows the wide availability range and high potential for high Reynolds number experiments provided by the NMIJ facilities.

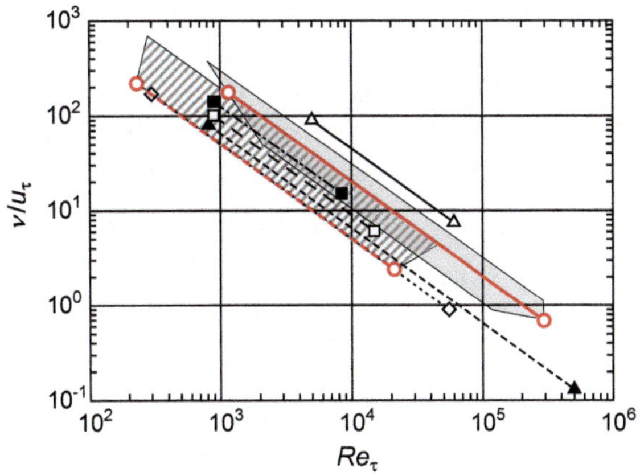

Fig. 5 Experimental/applicable range plotted against the viscous length scale and the friction Reynolds number. *White triangle* CICLoPE (D = 900 mm) [3], *black triangle* Superpipe (D = 129 mm) [2], *black rectangular* CoLaPipe (D = 190 mm) [6], *white rectangular* Laufer (D = 254 mm) [7], *Diamond* Nikuradse (D = 100 mm) [1], *bold circles* Furuichi (D = 100 and 387 mm) [5], *Gray shaded area* available range of Hi-Reff (D = 200–600 mm), *Diagonally hatched area* available range of the water flow facility (D = 100–400 mm)

3 Uncertainty for the Measurement of Friction Factor

In this section, an uncertainty analysis for the friction factor is presented to show the advantage provided by low uncertainty flow rate measurements. The relative standard uncertainty in the friction factor measurement is given by the following,

$$\frac{u(\lambda)}{\lambda} = \sqrt{\left\{\frac{u(\Delta p)}{\Delta p}\right\}^2 + \left\{5\frac{u(D)}{D}\right\}^2 + \left\{\frac{u(\rho)}{\rho}\right\}^2 + \left\{2\frac{u(q)}{q}\right\}^2 + \left\{\frac{u(L)}{L}\right\}^2}. \quad (1)$$

where, λ is the friction factor, Δp is the differential pressure, D is the pipe diameter, ρ is the density of water, q is the volumetric flow rate and L is the length between the two pressure taps. In this experiment, the differential pressure is measured as the average value from a set of four wall taps that are installed at the same streamwise position.

The uncertainty budget for the previous experiments [5] is shown in Table 1. In our previous experiments, pipes with different diameters ($D = 387$ and 100 mm) were used. The results clearly showed that the dominant uncertainty sources are the differential pressure and flow rate measurements. The uncertainty in the differential pressure measurement is primarily due to the uncertainty of the digital manometer and pressure deviation among the taps located at the same position. Here, it should be noted that the uncertainty contribution to the flow rate measurement is large, even though NMIJ facilities are capable of producing low uncertainty levels. This indicates, obviously, that the NMIJ facilities have a significant advantage in terms of

Table 1 Budget sheet for the measurement of friction factor

Uncertainty source	Uncertainty $D = 387$ mm (%)	Uncertainty $D = 100$ mm (%)
Differential pressure	0.401	0.401
Pipe diameter	0.162	0.125
Density of water	0.005	0.005
Flow rate	0.400	0.152
Tap distance	0.008	0.008
Expanded uncertainty ($k = 2$)	1.18	0.89

the accuracy when conducting turbulent pipe flow experiments in comparison with other facilities.

4 Summary

In this paper, details of the experimental facility at NMIJ, including Hi-Reff (which boasts a maximum Reynolds number of $Re_{\mathrm{D}} = 2.0 \times 10^7$) were reported. The most remarkable feature of the facility is its highly accurate 0.040–0.10% flow rate measurement capability, from which the highly accurate friction factor measurement (estimated to be 0.89–1.18%) has been achieved. From these factors, we can conclude that these NMIJ facilities have high potential for use as bases for turbulent pipe flow experiments.

References

1. J. Nikradse, Laws of turbulent flow in smooth pipe. NASA TT F-10 [English translation (1966)] (1932)
2. M.V. Zagarola, A.J. Smits, Mean flow scaling in turbulent pips flow. J. Fluid Mech. **373**, 33–79 (1998)
3. A. Talamelli, G. Bellani, A. Rossetti, The long pipe in CICLoPE: a design for detailed turbulence measurements. Prog. Turbul. V **149**, 131–142 (2014)
4. N. Furuichi, H. Sato, Y. Terao, M. Takamoto, A new calibration facility of flowrate for high Reynolds number. Flow Meas. Instrum. **20**(1), 38–47 (2009)
5. N. Furuichi, Y. Wada, Y. Terao, Y. Tsuji, Friction factor and mean velocity profile for pipe flow at high Reynolds number. Phys. Fluid **27**, 095108 (2015). doi:10.1063/1.4930987
6. F. König, E.S. Zanoun, E. Öngüner, C. Egbers, The CoLaPipe—the new Cottbus large pipe test facility at Brandenburg University of Technology Cottbus-Senftenberg. Rev. Sci. Instrum. **85**, 075115 (2014)
7. J. Laufer, The structure of turbulence in fully developed pipe flow. NACA report 1174 (1954)

Wavenumber Dependence of Very Large-Scale Motions in CICLoPE at $4800 \leq Re_\tau \leq 37{,}000$

Emir Öngüner, El-Sayed Zanoun, Tommaso Fiorini, Gabriele Bellani, Amir Shahirpour, Christoph Egbers and Alessandro Talamelli

Abstract The present work aims at investigating the very large-scale structures of turbulent pipe flow in CICLoPE at high Reynolds numbers. According to recent studies, some open questions remain to be answered to identify accurate sizes of these turbulent structures in pipe flow. The CICLoPE facility has been therefore utilized, providing an opportunity to approach high Reynolds number flows with high enough resolution in terms of the viscous length scale, allowing us to investigate the behavior of such turbulent structures. Meandering structures, usually referred as VLSM (very large-scale motions), have been identified with claimed extension up to $20R$, where R is the pipe radius.

1 Introduction

Characterization of turbulence structures in pipe flows at high Reynolds numbers are of vital importance with respect to the Reynolds stresses and the turbulent kinetic energy production. Recently, there has been an increasing interest in observing and understanding of the large organized vortex clusters forming large-scale and very large-scale motions (LSM and VLSM). Nevertheless, a solid definition of their nature and vivid understanding of their evolutions are still progressing. Therefore, this study is mainly focusing on clarifying the nature of the VLSM as well as describing and identifying them in quantitative manner.

E. Öngüner (✉) · E.-S. Zanoun · A. Shahirpour · C. Egbers
Department of Aerodynamics and Fluid Mechanics, Brandenburg University of Technology, 03046 Cottbus, Germany
e-mail: emir.oenguener@b-tu.de

T. Fiorini · G. Bellani · A. Talamelli
Department of Industrial Engineering, Università di Bologna, 47100 Forli, Italy

E.-S. Zanoun
Faculty of Engineering, Benha University, Benha 13512, Egypt

R. Örlü et al. (eds.), *Progress in Turbulence VII*, Springer Proceedings in Physics 196, DOI 10.1007/978-3-319-57934-4_15

Recent investigations of fully developed turbulent pipe flows, e.g. Bailey et al. [1] and Bosenberg et al. [2] showed that identification of the LSM and the VLSM in pipes are still under progressing and it might be concluded that there is still a lack of a complete and common definition for scales of those structures, in particular, at very high Reynolds numbers. Therefore, a quantitative measure of the Reynolds stresses and the energy production associated with such scales are indeed to be clearly defined.

An experimental attempt by Guala et al. [3] suggested a value of 2 for the normalized wavenumber $k_x R$ as a separator between the range of VLSMs and LSMs. It is clearly observable that according to the domain dividing method in premultiplied power spectra, the wavenumbers are converted to wavelengths independently. This is the reason why the results are shown separately for LSM and VLSM, in [3]. The line of separation is corresponding to a wavelength of πR, which is slightly longer than the accepted mean length of a turbulent bulge. They also found that the streamwise energetic modes can extend up to 12–14 pipe radii and 4–5 times longer than the length of a turbulent bulge and referred to these modes as very large-scale motions (VLSM). More recently, [4] used the whole premultiplied spectra and the observable peaks obtained to determine the wavenumbers for each radial location. The only difference to [3] was that structures were compared according to their dimensionless wavenumbers. On the other hand, [2] and [4] followed another method by associating the highest peaks in the power spectra to a possible turbulent structure.

2 Objectives and Measurements

The measurements in CICLoPE facility aim at investigating one-dimensional energy spectra over a wide range of Reynolds number. Considering similar facilities, Cottbus Large Pipe (CoLaPipe) [5, 6] covers a range of $10^3 \leq Re_\tau \leq 17 \times 10^3$ with large enough viscous length scales ($10 \leq \ell^* \leq 300$), and the Princeton/ONR Superpipe with approximately $3 \times 10^3 \leq Re_\tau \leq 10^5$ working range, however, having much smaller size of the viscous length scales ($0.5 \leq \ell^* \leq 30$). Here Re_τ is called the shear Reynolds number and defined as $Re_\tau = Ru_\tau/\nu$, where u_τ is the wall friction velocity, R is the pipe radius and ν is the kinematic viscosity of the fluid. The CICLoPE covers a region defined by $5 \times 10^3 \leq Re_\tau \leq 37 \times 10^3$ and sizes of viscous scales ($10 \leq \ell^* \leq 100$).

It is worth noting that the energy distribution in wavenumber domain helps to understand the behavior and structures of turbulence. In pre-multiplied spectra, the signature of the coherent motions can be easily observed where the spectral peaks, i.e. inner as well as outer spectral peaks, in near-wall and outer regions, respectively, provide evidences for the location of the maximum energy production and the energy content per wavenumber, see [4]. It was stated by [4] that the location, magnitude and wavelength of these spectral peaks are still open issues. Following the Kolmogorov scaling laws, i.e. k_x^{-1} and $k_x^{5/3}$ and their dependence on the Reynolds number provide also inputs about the universal behaviour of the energy spectrum in smaller and larger length scales at different wall-normal positions.

The mean pressure gradient along the CICLoPE pipe test section was monitored using pressure taps equally-spaced and distributed along the pipe section. Utilizing the mean pressure gradient obtained, the wall shear stress and consequently the wall friction velocity are determined. In order to obtain turbulence statistical quantities, instantaneous velocity field is resolved using thermal anemometer. The most common thermal anemometry used to obtain information on time-averaged and fluctuating velocities is the hot-wire anemometry (HWA), working in constant temperature mode. Further processing of the HWA time trace data acquired, i.e. the statistical and spectral quantities, were carried out in order to assess the behavior of the CICLoPE turbulent pipe flow. It is worth noting here that only naturally generated turbulence was considered in this piece of work.

In order to compare the CICLoPE data with those from the CoLaPipe, a common shear Reynolds number range $4{,}800 \leq Re_\tau \leq 17{,}000$ is considered for a viscous length scale approximately $\ell^* \leq 100$, see Fig. 1 in [5]. The main goal is to scale the turbulent spectra at different Reynolds numbers and wall-normal positions to identify the signs of any possible LSMs and VLSMs. Spectral peak locations at inner and outer regions were studied for each Reynolds number. Within this context the size effect of the three different facilities (incl. Superpipe) on the spatial development of turbulent structures can be clarified in further studies. Similar measurements have been already conducted by Princeton team [2, 4].

3 Results and Discussions

Figure 1 shows the premultiplied spectra normalized by the shear velocity ($k_x \Phi_{uu}/u_\tau^2$) at the lowest and highest Reynolds number cases in the CICLoPE facility. A good collapse is observed for $k_x R < 3$ (Fig. 1a), and for $k_x R < 2$ (Fig. 1b), in particular for $y/R < 0.3$ at $Re_\tau = 4800$ and $Re_\tau = 37{,}000$, respectively. The first peaks of premultiplied power spectra at various wall-normal locations seem to appear for similar normalized wavenumber. The first peak of spectra is considered as an evidence of the VLSM which is originating at lower wavenumbers. The magnitudes of the VLSM peaks are observed to decreas at higher wall-normal positions in good agreement with [4] which might be attributed to the decay of the total momentum transport as the wall distances increases beyond the outer limit of the inertial sublayer $y/R > 0.15$. In order to find the peaks of the energy spectra and estimate the wavenumber peak location the methodology adopted by [2] was followed. To reduce the scatter in the data a Gaussian window is applied to smooth the existing plots. Figure 1 shows significant differences in magnitudes of premultiplied spectra as observed in DNS results of [7] in comparison to [4]. CICLoPE delivers a viscous length scale ($\ell^* = \nu/u_\tau$) of 10.1 μm where Superpipe reaches only 1.7 μm at $Re_\tau = 37{,}000$. For this reason scaling

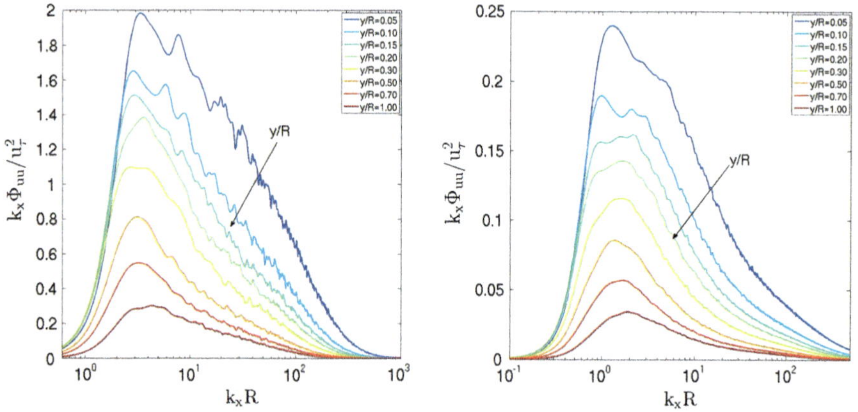

Fig. 1 Spectra with varying wall-normal distances at $Re_\tau = 4800$ (*left*) and $Re_\tau = 37{,}000$ (*right*)

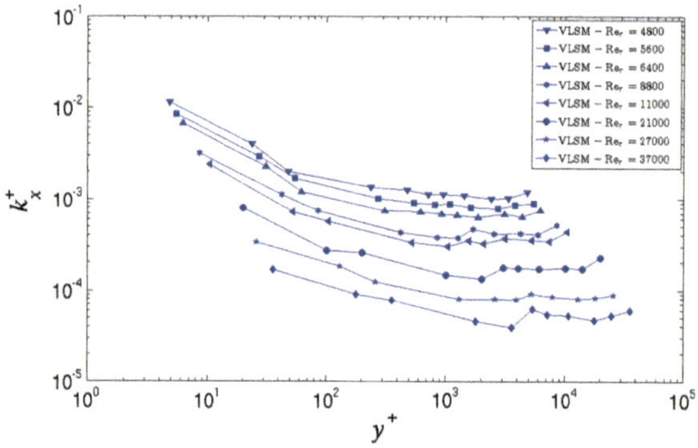

Fig. 2 Inner scaling of spectral peak locations of VLSMs for $4800 \leq Re_\tau \leq 37000$

of $k_x \Phi_{uu}$ with various u_τ in two facilities may deliver different magnitudes at the same shear Reynolds numbers.

Utilizing the viscous length scale ℓ^*, the wavenumber k_x was normalized yielding $k_x^+ = k_x \nu / u_\tau$ and then presented versus the normalized wall distance, ($y^+ = y u_\tau / \nu$) in Fig. 2. Figure 2 illustrates a monotonic decrease of the VLSM peaks up to the inner limit of the inertial sublayer ($y^+ \approx 3 \times 10^2$), reaching an almost asymptotic behavior along the inertial sublayer for various Reynolds numbers. The normalized wavenumber k_x^+ values corresponding to the same positions are shifted at higher y^+ values due to increasing the shear velocity with increasing the Reynolds number.

Spectral peak locations in outer coordinates, i.e. scaled with R, show an interesting trend in agreement with the Superpipe data described in [2]. The normalized wavenumber decreases monotonically along the wall distance $0.001 \leq y/R \leq 0.05$. On the other hand, at higher wall-normal distances, i.e. $y/R \leq 0.05$ which is close to the inner limit of the inertial sublayer, the normalized wavenumbers ($k_x R$) values are coming close to each other, and therefore, start to stabilize for all Re_τ till reaching the

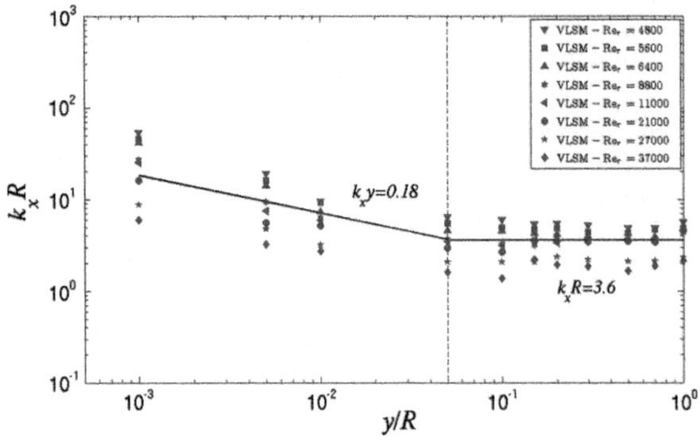

Fig. 3 Outer scaling of spectral peak locations of the VLSMs for $4800 \leq Re_\tau \leq 37000$

pipe centerline. One might observe from Fig. 3 that the VLSMs are about $k_x R \approx 3.6$ ($\lambda_x/R \approx 1.75$) which seem to be smaller than sizes expected according to [8] and [3]. In addition to different non-dimensionalized structure lengths, Figs. 2 and 3 show also some discrepancies according to results of [4]. One particular reason regarding the differences for inner and outer scaling can be addressed to unequal viscous conditions and different pipe radii. One should state that the inner diameter ratio of CICLoPE and Superpipe ($D_{CICLoPE}/D_{Superpipe} = 6.97 \approx 7$) is corresponding to the ratio of non-dimensionalizated wavelenghts of VLSMs in both facilities which are originating at similar wavenumbers: $(\lambda_x/R_{Superpipe})/(\lambda_x/R_{CICLoPE}) \approx 14/2 = 7$. It is obvious that a scaling of dimensionalized structure lengths with a smaller R will provide larger non-dimensionalized wavelengths than a larger R. For this reason higher $k_x R$ values in outer scaling in CICLoPE (Fig. 3) may lead to much smaller non-dimensionalized VLSM lengths in comparison to Superpipe and also [8] and [3].

4 Conclusions and Future Work

The current study confirms the importance of the very large-scale structures in understanding of high Reynolds number wall turbulence. The premultiplied spectra showed good collapse for $k_x R < 3$, in particular, for high enough Reynolds number. A monotonic decrease of the VLSM peaks within the wall layer was observed up to the inner limit of the inertial sublayer. On the contrary, an asymptotic behavior along the inertial sublayer was reached for various Reynolds numbers. Existing data of the Superpipe will be compared with CoLaPipe and CICLoPE in terms of their premultiplied spectra ($k_x \Phi_{uu}/u_\tau^2$) and wavenumber dependences of VLSM & LSM peaks. The normalization & scaling procedures of the turbulent structure lengths at different high Reynolds numbers pipe flow facilities with friction veloctiy (u_τ) and various diameters ($D_{Superpipe} = 0.129\,\text{m}$, $D_{CoLaPipe} = 0.19\,\text{m}$, $D_{CICLoPE} = 0.9\,\text{m}$)

will be reviewed again considering the smaller non-dimensionalized wavelengths obtained in CICLoPE than those reported earlier in literature.

Acknowledgements This work was supported under EuHIT (European High-Performance Infrastructures in Turbulence) grant with the title "Spectral scaling of turbulence in CICLoPE at high Reynolds numbers". The authors at Brandenburg University of Technology are supported by the DFG-German Research Foundation as a part of the FOR1182 and SPP1881 (grant nr.: EG100/24-1) project.

References

1. S.C.C. Bailey, M. Hultmark, A.J. Smits, M.P. Schultz, Azimuthal structure of turbulence in high Reynolds number pipe flow. J. Fluid Mech. **615**, 121–138 (2008)
2. B.J. Rosenberg, M. Hultmark, M. Vallikivi, S.C.C. Bailey, A.J. Smits, Turbulence spectra in smooth- and rough-wall pipe flow at extreme Reynolds numbers. J. Fluid Mech. **731**, 46–63 (2013)
3. M. Guala, S.E. Hommena, R.J. Adrian, Large-scale and very-large-scale motions in turbulent pipe flow. J. Fluid Mech. **554**, 521–542 (2006)
4. M. Vallikivi, B. Ganapathisubramani, A.J. Smits, Spectral scaling in boundary layers and pipes at very high Reynolds numbers. J. Fluid Mech. **771**, 303–326 (2015)
5. F. König, E.-S. Zanoun, E. Öngüner, C. Egbers, The CoLaPipe—the new Cottbus large pipe test facility at Brandenburg University of Technology Cottbus-Senftenberg. Rev. Sci. Instrum. **85**, 075115 (2014)
6. E.-S. Zanoun, E. Öngüner, C. Egbers, Conventional measuring probes in the wall layer of turbulent subsonic ducted flows. Thermophys. Aeromech. **23**(3), 329–342 (2016)
7. X. Wu, J.R. Baltzer, R.J. Adrian, Direct numerical simulation of a 30R long turbulent pipe flow at $Re_\tau = 685$: large- and very large-scale motions. J. Fluid Mech. **698**, 235–281 (2012)
8. K.C. Kim, R.J. Adrian, Very large-scale motion in the outer layer. Phys. Fluids **11**, 417–422 (1999)
9. A. Talamelli, F. Persiani, J.H.M. Fransson, A. Johansson, H.M. Nagib, J.D. Redi, K.R. Sreenivasan, P.A. Monkewitz, CICLoPE—a response to the need for high Reynolds number experiments. Fluid Dyn. Res. **41**(2), 021407 (2009)

Streamwise Auto-Correlation Analysis in Turbulent Pipe Flow Using Particle Image Velocimetry at High Reynolds Numbers

Emir Öngüner, El-Sayed Zanoun and Christoph Egbers

Abstract This preliminary study focuses on determining the lengths of turbulent pipe flow structures at $Re_b \approx 60,000$ and $140,000$ using auto-correlation analysis in streamwise direction considering velocity fluctuations, where Re_b is the Re-number based on bulk velocity. These structures are usually represented in terms of wavelengths λ or wavenumbers k. The current investigations on such turbulent structures including both large-scale motions (LSM) and very large-scale motions (VLSM) showed that the streamwise extension of these structures is highly dependent on the Reynolds number. Hence, the Cottbus large pipe (CoLaPipe) as a high Reynolds number test facility is being used to understand the physical processes and dynamics of such structures. These turbulence structures have been investigated using particle image velocimetry (PIV) to validate and compare earlier results obtained utilizing hot-wire anemometry (HWA).

1 Introduction

It has been recently observed that the sizes of turbulence structures can reach even $20R$ along the pipe axis, where R is the pipe radius, and are highly dependent on the Reynolds number [4, 5]. According to [1, 2, 6] some open questions remain unsettled for identifying accurate sizes of the large-scale motions (LSM) and very large-scale motions (VLSM). The theory of [2] claimed sizes of LSMs as $\lambda_{LSM} = 2R - 3R$ and VLSMs as $\lambda_{VLSM} = 8R - 16R$ using hot-wire anemometry at high frequencies in turbulent pipe flow. The estimated turbulent structure lengths in these studies have been calculated regarding one dimensional spectral analysis and temporal development. One should also consider that these investigations were based on similar pipe geometry; e.g. [2]: $D_i = 127\,\text{mm}$, [1]: $D_i = 127\,\text{mm}$, [6]: $D_i = 129\,\text{mm}$, where D_i

E. Öngüner (✉) · E.-S. Zanoun · C. Egbers
Department of Aerodynamics and Fluid Mechanics, Brandenburg University of Technology, 03046 Cottbus, Germany
e-mail: emir.oenguener@b-tu.de

E.-S. Zanoun
Faculty of Engineering, Benha University, 13512 Benha, Egypt

R. Örlü et al. (eds.), *Progress in Turbulence VII*, Springer Proceedings in Physics 196, DOI 10.1007/978-3-319-57934-4_16

is the inner diameter of pipe. The current study is targeting at using a non-intrusive measurement method, i.e. PIV, to scan a large domain of pipe section having larger diameter, $D_i = 190\,\text{mm}$, in fully developed turbulent state at $Re_b = 60,000$ and 140,000 to focus on spatial development of the large-scale structures.

2 Facility Description and Measurements

The CoLaPipe is providing an opportunity to approach higher Reynolds numbers, $60,000 \leq Re_b \leq 1,000,000$, with high enough resolution, $10 \leq \ell^* \leq 300$, where ℓ^* is the viscous length scale. The pipe test section is made out of acrylic glass, having inner pipe diameter of $D_i = 190 \pm 0.23$ mm with deviation less than 0.12%. The total length of the suction side, L = 28 m, provides a test section with a length-to-diameter ratio of $x/D_i \approx 148$. Further informations about the pipe facility and wall-layer measurements using conventional probes can be found in [3, 8].

Preliminary PIV measurements have been performed between x/D_i = 91-101 in collaboration with LaVision GmbH which covers 2 m of pipe test section, i.e. approximately a domain of 20R. 4 Imager sCMOS cameras with 2560 × 2160 pixels (provided by LaVision) with a standard Nd:YAG double-pulse laser with $f = 15$ Hz and $\lambda = 532$ nm have been used to cover this area with an acquisition frequency of $f = 15$ Hz (Fig. 1). All cameras have been synchronized to obtain simultaneous data. Calibration has been performed using a self-made calibration pattern which is inserted through the 2 m-pipe segment before the measurements. Standard fog fluid is used with aerosol generator to produce smoke particles. It should also be noted that to avoid any macro-cracks on acrylic glass surface, any kind of seeding materials with alcohol is to be renounced. For each case investigated, 100 snapshots have been taken. For post-processing only approx. 14R segment was taken into account due to stitching problems of the last camera. In terms of viscous wall unit, 64,276 mm distance between two vectors is delivering the first point at $y^+ \approx 110$, which can be considered as a low resolution study for this preliminary experiment.

3 Results

To apply auto-correlation function, the streamwise velocity fluctuations over 14R pipe segment at various wall-normal locations (y/R) were considered. For each y/R, all instantaneous velocity data were taken where the mean velocity of each y/R location is subtracted from these values. As seen in Fig. 2, which shows a correlation curve corresponding to only a single PIV snapshot, after applying the correlation method some periodic behaviors can be observed. The first point in the velocity domain is taken as a reference point and the neighbor points are correlated with this reference point. It can be easily observed from the figure that by moving

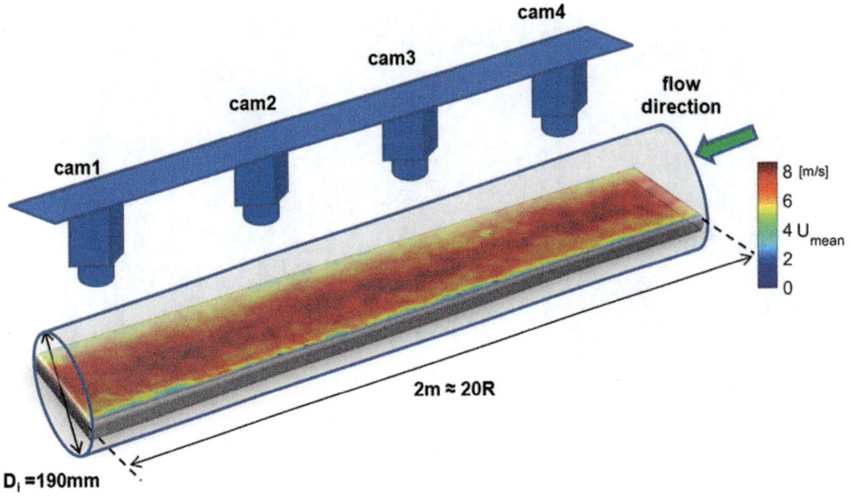

Fig. 1 Schematic of streamwise PIV arrangement in CoLaPipe using four sCMOS cameras, at $Re_b \approx 60,000$

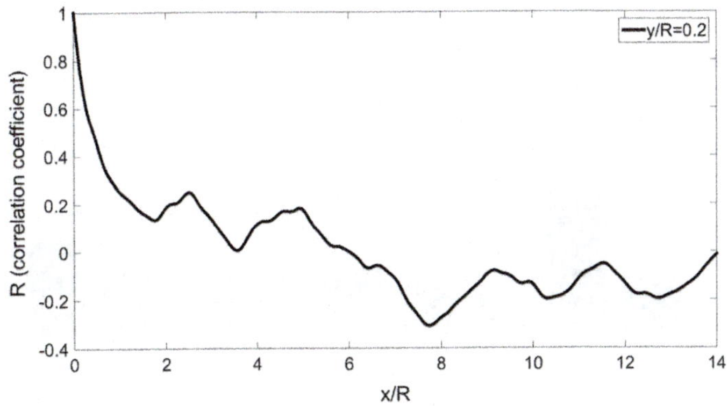

Fig. 2 The behavior of correlation coefficients over the 14R pipe domain at $Re_b \approx 60,000$ ($y/R = 0.2$) in a single PIV snapshot

in streamwise direction the correlation coefficient is decreasing but at some points repeated peaks can be distinguished.

A contour map illustrated in Fig. 3 exposes the variation of the streamwise correlations R_{uu} at $y/R = 0.2$ at lower Reynolds number case in a time-space plot. By checking all 100 streamwise PIV data, it should be noted that not all snapshots are delivering consistent velocity behavior which shows a periodic correlation. Figures 4 ($Re_b \approx 60,000$) and 5 ($Re_b \approx 140,000$) show the average distance between those correlation peaks (normalized by pipe radius) regarding all 100 snapshots. These repeated periodic distances are considered as signs of turbulent structures. A moving-average-function is applied to eliminate noises to capture the peaks properly. For

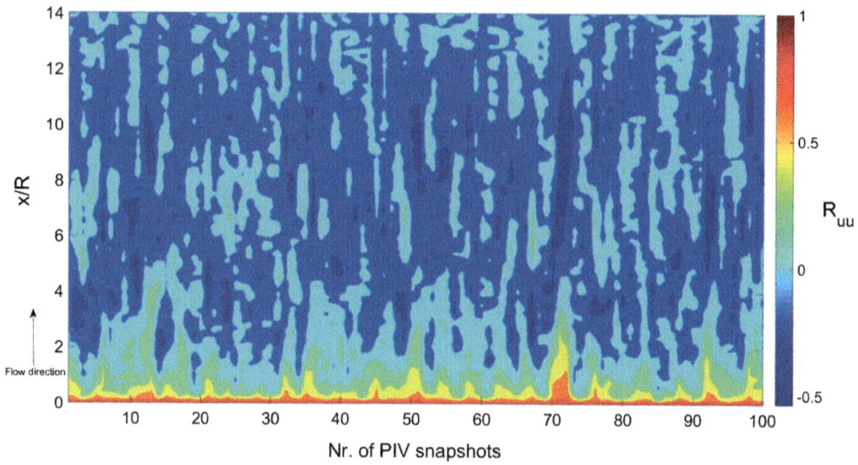

Fig. 3 Contour map showing variation of the streamwise correlation coefficients R_{uu} with 100 PIV-snapshots along 14R pipe domain at $y/R = 0.2$ and $Re_b \approx 60,000$

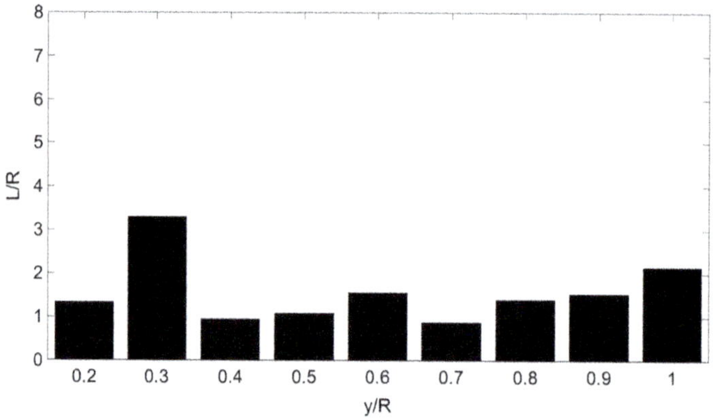

Fig. 4 Estimated mean structure lengths (L) normalized by R at $Re_b \approx 60,000$ of streamwise velocity fluctuations at various wall-normal locations over 100 snapshots

each snapshot the distances of visible peaks for a specific wall-normal location are calculated and the average values are presented in Figs. 4 and 5. According to this method, mean structure lengths of 1.6 and 2.5R can be observed for $Re_b \approx 60,000$ and $Re_b \approx 140,000$ respectively.

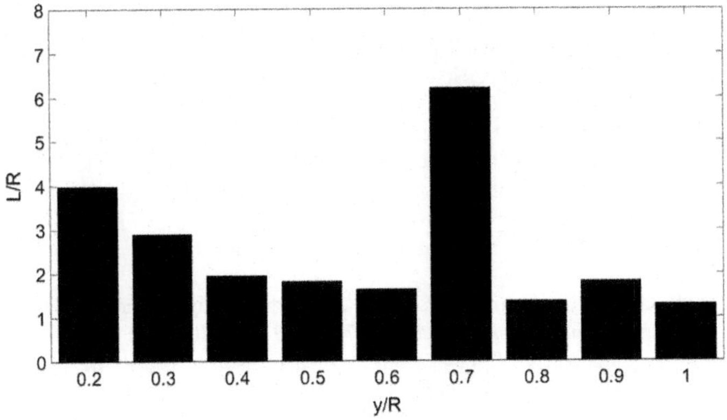

Fig. 5 Estimated mean structure lengths (L) normalized by R at $Re_b \approx 140,000$ of streamwise velocity fluctuations at various wall-normal locations over 100 snapshots

4 Future Work

High-speed PIV measurements with higher acquisition frequency and larger number of snapshots are essential to have a more precise analysis for determining the length of the structures in fully developed pipe flow. Alternatively, instead of using multi-point-correlation in a wall-normal line, applying a two-point-correlation in streamwise and spanwise directions will be implemented to investigate the spatial extend of correlated motions within the measurement plane. It should also be noted that the outer scaling factor R of the previous studies is widely used for facilities with similar pipe diameters. The effect of larger pipe diameters should be taken into account by normalization with R to determine the proper lengths of the large-scale structures.

Acknowledgements This work was supported by LaVision GmbH. We gratefully acknowledge many useful discussions with Mirko Dittmar and Peter Meyer. The authors are supported by the DFG-German Research Foundation as a part of the FOR1182 and SPP1881 (Grant No. EG100/24-1) project. CoLaPipe facility is used as a part of the EuHIT (European High-Performance Infrastructures in Turbulence) program.

References

1. M. Guala, S.E. Hommena, R.J. Adrian, Large-scale and very-large-scale motions in turbulent pipe flow. J. Fluid Mech. **554**, 521–542 (2006)
2. K.C. Kim, R.J. Adrian, Very large-scale motion in the outer layer. Phys. Fluids **11**, 417–422 (1999)
3. F. König, E.-S. Zanoun, E. Öngüner, C. Egbers, The CoLaPipe—the new Cottbus large pipe test facility at Brandenburg University of Technology Cottbus-Senftenberg. Rev. Sci. Instrum. **85**, 075115 (2014)

4. E. Öngüner, F. König, E.S. Zanoun, C. Egbers, Experimentelle Untersuchung der turbulenten Rohrstrukturen bei hohen Reynolds-Zahlen (2014) "Lasermethoden in der Strmungsmesstechnik", 9–11 Sept 2014, Karlsruhe, Hrsg.: Ruck, B., Gromke, C., Klausmann, K., Leder, A., Dopheide, D. ISBN 978-3-9816764-0-2
5. E. Öngüner, M. Dittmar, P. Meyer, C. Egbers, PIV-Messungen in horizontaler Rohrstrmung bei hohen Reynolds-Zahlen (2015) "Lasermethoden in der Strmungsmesstechnik", 8–10 Sept 2015, Dresden, Hrsg.: Czarske, J., Büttner, L., Fischer, A., Ruck, B., Leder, A., Dopheide, D. ISBN 978-3-9816764-1-9
6. M. Vallikivi, Wall-bounded turbulence at high Reynolds numbers. PhD Thesis, Princeton University (2014)
7. M. Vallikivi, B. Ganapathisubramani, A.J. Smits, Spectral scaling in boundary layers and pipes at very high Reynolds numbers. J. Fluid Mech. **771**, 303–326 (2015)
8. E.-S. Zanoun, E. Öngüner, C. Egbers, Conventional measuring probes in the wall layer of turbulent subsonic ducted flows. Thermophys. Aeromech. **23–3**, 329–342 (2016)

Part IV
Modelling

Subgrid-Scale Model with Structural Effects Incorporated Through the Helicity

Nobumitsu Yokoi and Akira Yoshizawa

Abstract By analogy with the Reynolds-averaged turbulence model, a subgrid-scale (SGS) model of turbulence with structural effect incorporated through the helicity is proposed. The SGS helicity implemented into the SGS stress is expected to suppress "too dissipative" properties of the Smagorinsky-type model applied to turbulence with coherent structures. Three types of helicity SGS models (two-, one-, and zero-equation models) are constructed depending on how many equations for the SGS statistical quantities should be solved in addition to the grid-scale (GS) momentum equation.

1 Introduction

Large-eddy simulation (LES) provides a powerful tool for investigating the real-world unsteady inhomogeneous turbulence at huge Reynolds numbers. Since the unresolved or subgrid-scale (SGS) components in LESs represent only small portion (smaller scales) of the flow components as compared with the turbulence of the Reynolds-averaged simulations, it is expected that the SGS model in the LESs can be much simpler than the counterpart in the Reynolds-averaged simulations. One of the simplest possible models is the eddy-viscosity type one, where the SGS stress $\tau^{ij} \equiv \overline{u^i u^j} - \overline{u}^i \overline{u}^j$ is expressed in terms of the grid-scale (GS) velocity strain $\overline{s}^{ij} (\equiv \partial \overline{u}^j / \partial x^i + \partial \overline{u}^i / \partial x^j)$ as

$$\tau_{\mathrm{D}}^{ij} = -\nu_{\mathrm{S}} \overline{s}_{\mathrm{D}}^{ij}, \tag{1}$$

N. Yokoi (✉)
Institute of Industrial Science, University of Tokyo, 4-6-1 Komaba, Meguro, Tokyo 153-8505, Japan
e-mail: nobyokoi@iis.u-tokyo.ac.jp

A. Yoshizawa
Emeritus at Institute of Industrial Science, University of Tokyo, Tokyo, Japan
e-mail: ay-tsch@mbg.nifty.com

R. Örlü et al. (eds.), *Progress in Turbulence VII*, Springer Proceedings in Physics 196, DOI 10.1007/978-3-319-57934-4_17

Table 1 Smagorinsky constant C_S and vortical structures

Flow types	C_S	Vorticity structure
Isotropic	0.18	Weak
Mixing layer	0.15	Intermediate
Wall	0.10	Strong

where $\overline{f}$ denotes the grid-scale (GS) or filtered component of f, and $A_{\rm D}^{ij} (\equiv A^{ij} - \delta^{ij} A^{\ell\ell}/3)$ is the deviatoric part of any tensor $\mathbf{A} = \{A^{ij}\}$. Here, $\nu_{\rm S}$ is the SGS viscosity, whose expression in the Smagorinsky model is given by

$$\nu_{\rm S} = (C_{\rm S} \varDelta)^2 \overline{s} \tag{2}$$

with the filter width $\varDelta$, the model constant $C_{\rm S}$, and $\overline{s} \equiv \sqrt{(\overline{s}^{ij})^2/2}$.

The classical Smagorinsky model is known to be "too dissipative" when it is applied to a turbulent flow with persistent flow structures. This is typically seen in situations where the Smagorinsky constant $C_{\rm S}$ has to be adjusted from flow to flow as shown in Table 1. It follows from (2) that the effective viscosity with the value of $C_{\rm S} = 0.18$ optimised for the homogeneous isotropic turbulence is more than three times too dissipative if it is applied to wall turbulence (where the optimised value is 0.10).

This need of constant adjustment is one of the major drawbacks of the classical Smagorinsky model for the SGS viscosity, as well as the lack of self-adaptation (namely, the SGS viscosity should vanish in the region where the flow is fully resolved, and decay correctly in the near wall region). These problems have been dissolved in the dynamic procedure, where the Smagorisky coefficient is dynamically estimated through the flow properties represented by the grid-scale (GS) velocity strain and with consecutive application of two filters. However, even with the dynamic procedure, it is recognised that the Smagorinsky model is too dissipative if it is applied to a flow with coherent structures, and implementation of such structure effects into the SGS model is important [1]. Actually, the stream-wise vorticity structures are ubiquitously observed not only in the near wall region but also in the mixing-layer turbulence with a large-scale velocity shear [2, 3] (Table 1).

This situation suggests that structural information of turbulence such as the stream-wise vorticity should be incorporated into the SGS modelling. Since such a vorticity in turbulence is directly related to the turbulent helicity, it is of importance to consider how to implement the effects of helicity into the SGS modelling.

2 Helicity

The local density of the turbulent helicity is defined by the correlation between the velocity fluctuation $\mathbf{u}'$ and vorticity fluctuation $\boldsymbol{\omega}'(= \nabla \times \mathbf{u}')$ as $H \equiv \langle \mathbf{u}' \cdot \boldsymbol{\omega}' \rangle (\langle \cdots \rangle$: ensemble average). The effects of helicity in inhomogeneous turbulence have been analytically investigated [4]. On the basis of the analytical Reynolds-stress expression, a turbulence model in terms of one-point statistical quantities was constructed as

$$\langle u'^i u'^j \rangle_{\mathrm{D}} = -\nu_{\mathrm{T}} S_{\mathrm{D}}^{ij} + \eta_{\mathrm{T}} \left[(\nabla H)^i \Omega_*^j + (\nabla H)^j \Omega_*^i \right]_{\mathrm{D}}, \tag{3}$$

where $S^{ij} (= \partial U^j / \partial x^i + \partial U^i / \partial x^j)$ is the mean velocity strain-rate tensor, and $\boldsymbol{\Omega}_* (= \boldsymbol{\Omega} + 2\boldsymbol{\omega}_{\mathrm{F}})$ is the mean absolute vorticity [$\boldsymbol{\Omega} (= \nabla \times \mathbf{U})$: mean relative vorticity, $\boldsymbol{\omega}_{\mathrm{F}}$: angular velocity]. The transport coefficients in (3) are modelled as

$$\nu_{\mathrm{T}} = K\tau = C_\nu K^2/\varepsilon, \quad \eta_{\mathrm{T}} = \ell^2 \tau = C_\eta K^4/\varepsilon^3, \tag{4}$$

where $K (= \langle \mathbf{u}'^2 \rangle / 2)$ is the turbulent energy, ε is its dissipation rate (τ: eddy turnover time, ℓ: length scale of energy-containing eddy, C_ν and C_η: model constants). The Reynolds-stress expression (3) indicates that the inhomogeneous helicity coupled with the mean absolute vorticity Ω_* may counter-balance with the eddy viscosity ν_{T}, and contribute to the sustainment of large-scale structures by effectively suppressing the enhanced momentum transport by turbulence [4]. Recently it has also been confirmed by direct numerical simulations (DNSs) of helically forced turbulence that a global flow can be generated by this inhomogeneous helicity effect [5].

3 Subgrid-Scale Modelling

With a filtering procedure, any field quantity f is divided into the grid-scale (GS) component $\overline{f}$ and subgrid-scale (SGS) component f'' as $f = \overline{f} + f''$. The equation of the GS velocity $\overline{\mathbf{u}}$ is written as

$$\left(\frac{\partial}{\partial t} + \overline{\mathbf{u}} \cdot \nabla \right) \overline{\mathbf{u}} = -\nabla \overline{p} - \nabla \cdot \boldsymbol{\tau} + \nu \nabla^2 \overline{\mathbf{u}}, \tag{5}$$

where $\overline{p}$ is the GS pressure and $\boldsymbol{\tau} = \{\tau^{ij}\}$ is the SGS stress tensor. In order to alleviate the problem of model-constant adjustment in the classical Smagorinsky model (1), we implement the effects of helicity into the SGS modelling. There have been some attempts to implement the effects of helicity into the SGS modelling of the helical turbulence [6, 7]. Here, we directly extend the expression for the Reynolds-stress expression (3) to the SGS stress as

$$\tau_{\rm D}^{ij} = -\nu_{\rm S}\overline{s}_{\rm D}^{ij} + (\Gamma_{\rm S}^{i}\overline{\omega}_{*}^{j} + \Gamma_{\rm S}^{j}\overline{\omega}_{*}^{i})_{\rm D} \tag{6a}$$

$$= -\nu_{\rm S}\overline{s}_{\rm D}^{ij} + \eta_{\rm S}[(\nabla H_{\rm S})^{i}\overline{\omega}_{*}^{j} + (\nabla H_{\rm S})^{j}\overline{\omega}_{*}^{i}]_{\rm D}, \tag{6b}$$

where $\overline{\boldsymbol{\omega}}_{*}(= \overline{\boldsymbol{\omega}} + 2\boldsymbol{\omega}_{\rm F})$ is the GS absolute vorticity.

On the analogy of the Reynolds-averaged model, the transport coefficients associated with $\overline{\mathbf{s}} = \{\overline{s}^{ij}\}$ and $\overline{\boldsymbol{\omega}_{*}}$, $\nu_{\rm S}$ and $\boldsymbol{\Gamma}_{\rm S}$, should be expressed in terms of the SGS energy and helicity, respectively. Depending on how to express them, the models are classified into three categories, two-, one-, and zero-equation models.

Two-Equation Model

The most straightforward approach is to introduce the SGS energy and helicity, $K_{\rm S}$ and $H_{\rm S}$, defined by

$$K_{\rm S} = (\overline{\mathbf{u}^2} - \overline{\mathbf{u}}^2)/2, \tag{7}$$

$$H_{\rm S} = \overline{\mathbf{u}\cdot\boldsymbol{\omega}} - \overline{\mathbf{u}}\cdot\overline{\boldsymbol{\omega}}, \tag{8}$$

and examine the spatiotemporal evolution of them. These quantities represent the statistical properties of unresolved or SGS motions. The transport coefficients in the SGS stress, $\nu_{\rm S}$ and $\boldsymbol{\Gamma}_{\rm S}$, are expressed in terms of $K_{\rm S}$ and $H_{\rm S}$ as well as Δ (the largest scale of the unresolved scales of motion): $\nu_{\rm S} = \nu_{\rm S}\{\Delta, K_{\rm S}\}$ and $\Gamma_{\rm S} = \Gamma_{\rm S}\{\Delta, K_{\rm S}, H_{\rm S}\} = \eta_{\rm S}\{\Delta, K_{\rm S}\}\nabla H_{\rm S}$. And the SGS energy dissipation rate $\varepsilon_{\rm S}$ is expressed also in terms of Δ and $K_{\rm S}$ as $\varepsilon_{\rm S} = \varepsilon_{\rm S}\{\Delta, K_{\rm S}\}$. In this case, the transport coefficients in (6) are expressed as

$$\nu_{\rm S} = C_{\rm S}\Delta K_{\rm S}^{1/2},\quad \eta_{\rm S} = C_{\eta\rm S}\Delta^2(\Delta/K_{\rm S}^{1/2}) = C_{\eta\rm S}\Delta^3 K_{\rm S}^{-1/2}. \tag{9}$$

In addition to the GS momentum equation (5) with (6), we solve the equations of the SGS energy $K_{\rm S}$ and the SGS helicity $H_{\rm S}$. The $K_{\rm S}$ equation is given by

$$\left(\frac{\partial}{\partial t} + \overline{\mathbf{u}}\cdot\nabla\right) K_{\rm S} = -\tau^{ab}\frac{\partial \overline{u}^b}{\partial x^a} - \varepsilon_{\rm S} + \nabla\cdot\mathbf{T}_{\rm S} \tag{10}$$

with the dissipation rate $\varepsilon_{\rm S}$ and the transport rate $\mathbf{T}_{\rm S}$ of the SGS energy defined by

$$\varepsilon_{\rm S} = C_{\varepsilon\rm S}K_{\rm S}^{3/2}/\Delta,\quad \mathbf{T}_{\rm S} = (\nu_{\rm S}/\sigma_{\rm S})\nabla K_{\rm S}, \tag{11}$$

where $C_{\varepsilon\rm S}$ and $\sigma_{\rm S}$ are model constants. On the other hand, the equation of the SGS helicity is given by

$$\left(\frac{\partial}{\partial t} + \overline{\mathbf{u}}\cdot\nabla\right) H_{\rm S} = -\tau^{ab}\frac{\partial \overline{\omega}^b}{\partial x^a} + \overline{\omega}^a\frac{\partial \tau^{ab}}{\partial x^b} - \varepsilon_{\rm HS} + \nabla\cdot(K_{\rm S}\overline{\boldsymbol{\omega}} + \mathbf{T}_{\rm HS}), \tag{12}$$

where the dissipation and transport rates of $H_{\rm S}$, $\varepsilon_{\rm HS}$ and $\mathbf{T}_{\rm HS}$, are modelled as

$$\varepsilon_{\mathrm{HS}} = C_{\varepsilon\mathrm{H}} H_{\mathrm{S}}/(K_{\mathrm{S}}/\varepsilon_{\mathrm{S}}) = C_{\varepsilon\mathrm{H}} \varepsilon_{\mathrm{S}} H_{\mathrm{S}}/K_{\mathrm{S}}, \quad \mathbf{T}_{\mathrm{HS}} = (\nu_{\mathrm{S}}/\sigma_{\mathrm{HS}}) \nabla H_{\mathrm{S}} \tag{13}$$

with $C_{\varepsilon\mathrm{H}}$ and σ_{HS} being the model constants.

In the framework of the two-equation model, two transport equations of the SGS quantities, K_{S} (10) and H_{S} (12), have to be simultaneously solved with the GS momentum equation (5).

One-Equation Model

In order to construct a simpler model, however, the number of the equations need to be reduced with some additional assumptions. We retain H_{S} as a SGS field quantity and solve the H_{S} equation since the evolution of SGS helicity is less known than the counterpart of the SGS energy. On the contrary, the K_{S} equation is simplified by assuming a local equilibrium between the production and dissipation rates of K_{S}. All quantities are to be expressed in terms of Δ, $\overline{s}$ and H_{S}, without resorting to K_{S}. In the framework of one-equation model, we use $\varepsilon_{\mathrm{S}} = \varepsilon_{\mathrm{S}}\{\Delta, \overline{s}\}$, $\nu_{\mathrm{S}} = \nu_{\mathrm{S}}\{\Delta, \overline{s}\}$, and $\boldsymbol{\Gamma}_{\mathrm{S}} = \boldsymbol{\Gamma}_{\mathrm{S}}\{\Delta, \overline{s}, H_{\mathrm{S}}\} = \eta_{\mathrm{S}}\{\Delta, \overline{s}\} \nabla H_{\mathrm{S}}$.

We assume the local equilibrium of the SGS energy, i.e., the production rate of K_{S} is balanced by the dissipation rate ε_{S} in (10) as

$$-\tau^{ij} \frac{\partial \overline{u}^j}{\partial x^i} \simeq \varepsilon_{\mathrm{S}}. \tag{14}$$

We approximate the SGS stress τ^{ij} by (1) and the SGS energy dissipation rate ε_{S} by (11), with eliminating K_{S} by (9). Then (14) reads

$$\nu_{\mathrm{S}} \overline{s}^2 = (C_{\varepsilon\mathrm{S}}/C_{\nu\mathrm{S}}^3) \nu_{\mathrm{S}}^3/\Delta^4, \tag{15}$$

which immediately gives us the Smagorinsky model of the SGS viscosity as (2) with $C_{\mathrm{S}} = (C_{\nu\mathrm{S}}^3/C_{\varepsilon\mathrm{S}})^{1/4}$. In this case, the SGS energy K_{S}, its dissipation rate, ε_{S}, and the helicity-related coefficient η_{S} are expressed as

$$K_{\mathrm{S}} = (\nu_{\mathrm{S}}/C_{\nu\mathrm{S}}\Delta)^2 = C_{\mathrm{KS}} \Delta^2 \overline{s}^2, \tag{16}$$

$$\varepsilon_{\mathrm{S}} \simeq -\tau^{ij} \frac{\partial \overline{u}^j}{\partial x^i} \simeq \nu_{\mathrm{S}} \overline{s}^2 = (C_{\mathrm{S}}\Delta)^2 \overline{s}^3, \tag{17}$$

$$\eta_{\mathrm{S}} = C_{\eta\mathrm{S}} \Delta^3 K_{\mathrm{S}}^{-1/2} = C_{\mathrm{HS}} \Delta^2/\overline{s} \tag{18}$$

with $C_{\mathrm{KS}} = (C_{\varepsilon\mathrm{S}} C_{\nu\mathrm{S}})^{-1/2}$ and $C_{\mathrm{HS}} = C_{\eta\mathrm{S}} C_{\varepsilon\mathrm{S}}^{1/2}/C_{\nu\mathrm{S}}^{1/2}$.

In this one-equation model, in addition to the GS momentum equation (5), the SGS helicity equation (12) should be solved with $\mathbf{T}_{\mathrm{HS}}$ (13) and

$$\varepsilon_{\mathrm{HS}} = C_{\varepsilon\mathrm{H}} H_{\mathrm{S}}/\overline{s}^{-1} = C_{\varepsilon\mathrm{H}} H_{\mathrm{S}} \overline{s}. \tag{19}$$

Zero-Equation Model

The Smagorinsky model is one of the simplest possible SGS models, which has no additional equations other than the GS momentum equation (zero-equation model). In order to construct a model as simple as the Smagorinsky model, we further assume the local equilibrium of the SGS helicity, i.e., the production rate of the SGS helicity is balanced with its dissipation rate $\varepsilon_{\rm HS}$. In this case, from (12) we have

$$- \tau^{ab} \frac{\partial \overline{\omega}^b}{\partial x^a} + \overline{\omega}^a \frac{\partial \tau^{ab}}{\partial x^b} \simeq \varepsilon_{\rm HS}. \tag{20}$$

Under this local equilibrium condition for the SGS helicity production rate with the SGS helicity dissipation rate (13), $H_{\rm S}$ can be estimated as

$$H_{\rm S} = \frac{1}{C_{\varepsilon {\rm H}} \overline{s}} \varepsilon_{\rm HS} \simeq \left[\nu_{\rm S} \overline{s}^{ij} \frac{\partial \overline{\omega}^j}{\partial x^i} - \overline{\omega}^i \frac{\partial}{\partial x^j} \nu_{\rm S} \overline{s}^{ij} + \frac{2}{3} \left(\overline{\boldsymbol{\omega}} \cdot \nabla \right) K_{\rm S} \right], \tag{21}$$

where the SGS stress (6) has been used with $K_{\rm S}$ (16), $H_{\rm S}$ (21), $\nu_{\rm S}$ (2), and $\eta_{\rm S}$ (18).

4 Discussion

We proposed a SGS model with the structural effects incorporated through the SGS helicity. In this model, effects of coherent structures in turbulence are taken into account through the terms of the inhomogeneous helicity which couples with the GS absolute vorticity in (6). If these terms work as a counter-balancer to the SGS viscosity term, as was already confirmed in the context of the Reynolds-averaged model, "too dissipative" properties of the Smagorinsky model might be alleviated. In this scenario, the Smagorinsky constant should be fixed to $C_{\rm S} = 0.18$ (optimised for isotropic turbulence) for any kind of turbulent flows, and the effects of vortical structures would effectively suppress the momentum transport without resorting to any constant adjustment. The validation of this scenario can be accomplished only through applications of the model to several kinds of turbulent flows.

It is worthwhile for remarking the following point. In the present zero-equation model, the SGS helicity is expressed in terms of the GS quantities as in (21). This estimate is based on the local equilibrium assumption (20), which is not necessarily the case depending on the flow configurations. The validity of (21) is under investigation with the aid of direct numerical simulations (DNSs) of a rotating stratified turbulence. If the estimate (21) is reasonably well, we can use the zero-equation model. If not, we should use the one- or two-equation model.

Acknowledgements Numerical simulations for the validation of the SGS model are performed on the XC-30 of Center for Computational Astrophysics (CfCA), National Astronomical Observatory of Japan (NAOJ). Part of this work was supported by the JSPF Bilateral Project between Japan and Russia.

References

1. H. Kobayashi, The subgrid-scale models based on coherent structures for rotating homogeneous turbulence and turbulent channel flow. Phys. Fluids **17**, 045104-1–045104-12 (2005)
2. S.K. Robinson, Coherent motions in the turbulent boundary layer. Annu. Rev. Fluid Mech. **23**, 601–639 (1991)
3. J.M. Wallace, Highlights from 50 years of turbulent boundary layer research. Annu. Rev. Fluid Mech. **13**, 53-1–53-70 (2013)
4. N. Yokoi, A. Yoshizawa, Statistical analysis of effects of helicity in inhomogeneous turbulence. Phys. Fluids A **5**, 464–477 (1993)
5. N. Yokoi, A. Brandenburg, Large-scale flow generation by inhomogeneous helicity. Phys. Rev. E **93**, 033125-1–033125-14 (2016)
6. Y. Li, C. Meneveau, S. Chen, G.L. Eyink, Subgrid-scale modeling of helicity and energy dissipation in helical turbulence. Phys. Rev. E **74**, 026310-1–026310-13 (2006)
7. C. Yu, Z. Xiao, Y. Shi, S. Chen, Joint-constraint model for large-eddy simulation of helical turbulence. Phys. Rev. E **89**, 043021-1–043021-12 (2014)

Fractional Turbulence Models

Peter W. Egolf and Kolumban Hutter

Abstract In this article we propose to generalize Reynolds shear stresses in local zero-equation turbulence models to nonlocal and fractional forms. In the well-accepted general method, starting with a Kraichnanian convolution-integral as Reynolds shear stress, different weighting functions are possible candidates to serve this purpose; e.g. the Liouville weighting function leads to the left-handed Riemann–Liouville fractional derivative and the Heaviside distribution to a mean velocity difference, respectively a difference quotient. Therefore, this weighting function transforms the first gradient (the one in the eddy diffusivity) of Prandtl's 1925 mixing-length model to an eddy diffusivity with a mean velocity difference and, thereby, directly leads to the (modified) Prandtl shear-layer model of 1942. Prandtl's intuitive development—which is in agreement with fractional calculus—does not serve as a proof of correctness, but is a welcome coincidence. By further following Prandtl's intuition and applying the Heaviside distribution also to the remaining driving gradient, yields the *Difference-Quotient Turbulence Model* (*DQTM*), which was discovered by other means and had been published by Egolf in 1991. As a result, it becomes clear that the DQTM is a natural nonlocal extension of Prandtl's models and contains a special case of a simple fractional derivative, namely a difference quotient, which stands for the highest possible nonlocality and minimum calculation time to solve a turbulent flow problem.

1 A Brief Introduction to Fractional Dynamics

Fractional calculus is attributed to De l'Hôpital, Leibniz, Riemann and Liouville (see e.g. [1]) with its roots going back to the end of the 17th century. For a long period of time it was exclusively attracting pure mathematicians to study fractional

P.W. Egolf (✉)
University of Applied Sciences of Western Switzerland, Yverdon, Switzerland
e-mail: Peter.egolf@heig-vd.ch

K. Hutter
Swiss Federal Institute of Technology, ETH, Zurich, Switzerland
e-mail: hutter@vaw.baug.ethz.ch

R. Örlü et al. (eds.), *Progress in Turbulence VII*, Springer Proceedings in Physics 196, DOI 10.1007/978-3-319-57934-4_18

differentiation, integration and differential equations [1, 2]. With the development of fractal geometry by Richardson and Mandelbrot [3], also fractal dynamics and thermodynamics [4] was developed and e.g. the connections between fractal dimensions and the orders of fractional derivatives were revealed [5]. Furthermore, in the last three decades, it was discovered that fractional calculus is the adequate tool to solve *complex problems* with memory effects and nonlocalities, scale-free self-similar processes, systems with non-Gaussian and long-range statistics (e.g. Lévy walks), non-Newtonian dynamics, non-equilibrium thermodynamics and dynamical phase transitions. Einstein demonstrated that Brownian motion leads straight-forwardly to a linear diffusion equation. Therefore, it is not astonishing that Lévy statistics, which is also adequate to model intermittency in turbulence, produces an anomalous diffusion equation with fractional derivatives [5]. From this point of view, it becomes clear that turbulent eddy diffusion may be successfully modeled by *fractional Reynolds stresses* [6].

2 A Theoretical Prologue

Turbulent flows fluctuate in space and time. This fact led scientists already at the end of the nineteenth century to apply statistical methods to describe 'turbulence'. Reynolds' [7] idea was to decompose a physical quantity ξ into an average $\langle\xi\rangle$ and its fluctuation ξ',

$$\xi = \langle\xi\rangle + \xi'. \tag{1}$$

Substituting all physical quantities, in the decomposition of (1), into the Navier–Stokes equations (NS) and additionally averaging the full equation, on the basis that turbulent motions are quasi-steady chaotic processes, the Reynolds-Averaged Navier–Stokes (RANS) equations for incompressible fluids are derived (see [8, 9])

$$\frac{\partial\langle\vec{v}\rangle}{\partial t} + \mathrm{div}(\langle\vec{v}\rangle \otimes \langle\vec{v}\rangle) + \frac{1}{\rho}\mathrm{grad}\langle p\rangle - \frac{1}{\rho}\mathrm{div}\langle\mathbf{T}_D\rangle + \frac{1}{\rho}\mathrm{div}\langle\mathbf{T}_R\rangle = 0, \quad \mathrm{div}\langle\vec{v}\rangle = 0, \tag{2a,b}$$

where $\vec{v} = (u_1, u_2, u_3)$ denotes the 3-d velocity field, p the scalar pressure field, $\langle\mathbf{T}_D\rangle$ the averaged dissipation tensor and $\langle\mathbf{T}_R\rangle$ a tensor, created by the averaging process, called Reynolds stress tensor. In Cartesian components, and by now denoting the average of the nine occurring scalar tensor elements by an overbar, $\langle\mathbf{T}_R\rangle$ takes the form (see e.g. [8])

$$\langle\mathbf{T}_R\rangle = \langle\vec{v}' \otimes \vec{v}'\rangle = \begin{pmatrix} \overline{u_1'^2} & \overline{u_1'u_2'} & \overline{u_1'u_3'} \\ \overline{u_2'u_1'} & \overline{u_2'^2} & \overline{u_2'u_3'} \\ \overline{u_3'u_1'} & \overline{u_3'u_2'} & \overline{u_3'^2} \end{pmatrix}. \tag{3a,b}$$

Because this tensor is symmetric, six new unknown second-order correlations appear in the well-defined problem, without delivering transparently six additional equations. For further insight in the methods of solving this closure problem, consult e.g. work by Kraichnan [10]. Here we concentrate only on the lowest-order method, namely that of describing second-order by first-order moments. This procedure closes the system of equations. The analogy between the tensors $\langle \mathbf{T}_D \rangle$ and $\langle \mathbf{T}_R \rangle$ led Saint Venant, Boussinesq and Prandtl, in analogy to the molecular kinetic theory of ideal gases and diffusion theory, to make the following local Fick-type parameterization [11]

$$\tau_{21} = -\rho \overline{u_2' u_1'} = \varepsilon_m \frac{\partial \bar{u}_1}{\partial x_2}. \tag{4a,b}$$

Notice that for a simple shear flow, with streamwise coordinate x_1 and spanwise coordinate x_2, this single second-order correlation is the only one of importance for the description of e.g. simple turbulent shear, Couette and Poiseuille flows. The quantity ε_m (index m for momentum) is not a material property; it is a model parameter that may depend on space and velocity coordinates, called *eddy diffusivity*.

3 Prandtl's Mixing-Length and Shear-Layer Model

The eddy diffusivities of *Prandtl's mixing-length model* of 1925 [11] and of his *shear-layer model* of 1942 [12] are

$$\varepsilon_m = \rho \sigma \chi_2^2 \left| \frac{\partial \bar{u}_1}{\partial x_2} \right|, \quad \varepsilon_m = \rho \sigma \chi_2 (\bar{u}_{1\,\max} - \bar{u}_{1\,\min}). \tag{5a,b}$$

The quantity σ is a constant and χ_2 is a (large) characteristic length of the flow in the flow-transverse x_2-direction. The {min, max}-velocities are defined by the equations

$$\bar{u}_{1\alpha}(x_1) = \underset{x_2}{\alpha} \{\bar{u}_1(x_1, x_2)\} = \bar{u}_1(x_1, x_{2\alpha}), \quad \alpha \in \{\min, \max\}, \tag{6a–d}$$

in which $x_{2\,\min}$ and $x_{2\,\max}$ are implicitly defined as those x_2-positions at fixed x_1, where $\bar{u}_1$ reaches a minimum and maximum, respectively.

By removing in (5b) the index *max*, the *modified Prandtl shear-layer model* [6] is obtained, which shows the following now also x_2-coordinate dependent eddy diffusivity

$$\varepsilon_m = \rho \sigma \chi_2 (\bar{u}_1 - \bar{u}_{1\,\min}) \tag{7}$$

and is a substantial improvement compared to the eddy diffusivities (5a,b). It is well known that the eddy diffusivity must vanish toward the edge of a turbulent domain, e.g. toward a wall, where the no-slip boundary condition requires that $\bar{u}_1 = 0$, a condition which is now fulfilled, because there $\bar{u}_1 = \bar{u}_{1\,\min}$.

4 A Recipe to Transform Local to Nonlocal Operators

The concept of generalizing nonlocally the derivative operators in (4b) or/and (5a) is now briefly outlined by presenting some results reported by Herrmann [13]. Let us consider a position of a mass point object in space described by $x(t)$. To this end, the velocity defined by the (Weierstrass) derivative is modified to be two-fold, namely

$$u(t) = \frac{dx}{dt} = \lim_{\tau \to 0} \frac{x(t) - x(t \pm \tau)}{\tau}, \tag{8a,b}$$

where $0 \leq \tau < \varepsilon$ and ε is a small positive real number. It is necessary to have information of the position of this object at a time $t - \tau$ or $t + \tau$, respectively. In this definition $\tau < \varepsilon$, $\varepsilon \to 0$. Therefore, the velocity is an *epsilon- or zero-nonlocal quantity*, respectively, or simply a *local quantity*. Even if for mathematically sufficiently continuous functions the two definitions lead to the same quantitative result, the first definition requires information backward in time (or in the past) and the second one forward in time (or in the future). Notice that the second case violates causality.

It is also possible to define the derivative by an integral operator

$$u(t) = \frac{d}{dt}x(t) = 2\int_0^{\infty} \delta(\tau)u\,(t-\tau)\,d\tau, \quad \int_{-\infty}^{\infty} \delta(t)f(t)dt = f\,(0)\,, \tag{9a–c}$$

where, because of causality reason, we restrict ourselves only to the backward derivative. The symbol $\delta(t)$ denotes the Dirac distribution (generalized function) with the property (9c). For a definition of the Dirac distribution (see e.g. Gelfand and Shilov [14]) numerous functions $\omega\,(\lambda, \tau)$ are candidates, as long as they are everywhere differentiable and also obey the following power-law attenuation requirement: $\lim_{t \to \infty} \omega\,(\lambda, t) = O\left(|t|^{-n}\right), \forall n > 0$. These functions are also called test functions. With the argument $|t|\,/\lambda$ ideal candidates are, for example, the exponential, the Airy, the sine function, etc. (see Table 1). Then it follows for a smooth parameter $\lambda \geq 0$, independent of the particular test function, that

$$\delta\,(t) = \lim_{\lambda \to 0} \omega\,(\lambda, t)\,. \tag{10}$$

In a straightforward manner, we substitute (10) into (9b) to obtain

$$\frac{d}{dt}x(t) = 2\lim_{\lambda \to 0}\int_0^{\infty} \omega\,(\lambda, t)\,u\,(t-\tau)\,d\tau. \tag{11}$$

Table 1 Weighting functions listed by their name, definition ω and norm Ω (a selection from [13]). In this article the two gray underlaid weighting functions are proposed to be applied

Name	Weight $\omega(\lambda, \tau)$	Norm Ω	Name	Weight $\omega(\lambda, \tau)$	Norm Ω
Inverse parabolic	$\frac{1}{1+(\tau/\lambda)^2}$	$\frac{\lambda\pi}{2}$	Sine	$\frac{\sin(\tau/\lambda)}{\tau}$	$\frac{\pi}{2}$
Airy	$\mathrm{Ai}\,(\tau/\lambda)$	$\frac{\lambda}{3}$	Caputo	$\tau^{\lambda-1}\theta\,(\lambda-\tau)$	$\Gamma\,(\lambda)$
Exponential	$e^{-(\tau/\lambda)}$	λ	Liouville	$\tau^{\lambda-1}$	$\Gamma\,(\lambda)$
Gaussian	$e^{-(\tau/\lambda)^2}$	$\lambda\Gamma\,(3/2)$	Heaviside (const.)	$\theta\,(\lambda-\tau)$	λ

Ai is the Airy and Γ the Gamma function

At this point our intention is to introduce higher nonlocality by dismissing the limit $\lambda \to 0$. In this generalization, we no longer require (10) to be fulfilled. With these generalizations, a nonlocal derivative is obtained

$$\hat{O}_{nonlocal}\,[f\,(t)] = \frac{1}{\Omega}\int_0^\infty \omega\,(\lambda, \tau)\,\hat{O}_{local}\left[\hat{s}_-\,(\tau)\,f\,(t)\right]d\tau, \tag{12a–c}$$

$$\hat{s}_-\,[f\,(t)] = f\,(t-\tau)\,,\quad \frac{1}{\Omega}\int_0^\infty \omega\,(\lambda, \tau)d\tau =: 1,$$

where the first-order derivative was replaced by a more general operator $\hat{O}$, and $\hat{s}_-$ denotes the negative shift operator, obeying the rule (12b), and Ω the normalization factor defined by (12c).

Notice that not all operators commute with the shift operator. Equation (12a) is a clear instruction how from a local a nonlocal operator can be created, namely by applying four rules R1–R5 (see also in [13]):

(R1) Starting with the local operator $\hat{O}_{local}$,
(R2) Choosing the appropriate shift operator, in our case $\hat{s}_-(\tau)$,
(R3) Choosing the appropriate weighting function ω *with its corresponding norm* Ω,
(R4) To produce the nonlocal convolution operator
(R5) To integrate over the appropriate domain.

5 Nonlocalization of Operators and Nonlocal Turbulence Models

The rules (R1)–(R4) in Sect. 4 are an efficient tool to generalize local turbulence models to different nonlocal counterparts. The entire method shall now be applied in some examples, a pure mathematical one and four physical examples considering nonlocal turbulence modeling.

5.1 *Liouville Fractional Derivative*

We choose a commuted Eq. (12a) that is also presented in [13]

$$\hat{O}_{nonlocal}\left[f(t)\right] = \frac{1}{\Omega}\hat{O}_{local}\left[\int\limits_0^\infty \omega(\lambda,\tau)\,\hat{s}_-(\tau)\,f(t)\,d\tau\right] \tag{13}$$

and consider the Liouville weight $\omega(\lambda,\tau) = \tau^{\lambda-1}(0 < \lambda < 1)$, norm $\Omega = \Gamma(\lambda)$ (see Table 1), the first-derivative and the negative shift operator $\hat{s}_-$. This transforms (13) to

$$\left.\frac{d}{dt}\right|_{nonlocal} f(t) = \frac{1}{\Gamma(\lambda)}\left.\frac{d}{dt}\right|_{local}\int\limits_0^\infty \tau^{\lambda-1} f(t-\tau)\,d\tau\,. \tag{14}$$

By substituting $\lambda = 1-\alpha$ and $\xi = t-\tau$, it follows that

$$\left.\frac{d}{dt}\right|_{nonlocal} f(t) = \frac{1}{\Gamma(1-\alpha)}\left.\frac{d}{dt}\right|_{local}\int\limits_{-\infty}^{t} (t-\xi)^{-\alpha} f(\xi)\,d\xi =: {}_L D_+^{\alpha} f(t). \tag{15}$$

${}_L D_+^{\alpha} f(t)$ is the *left-handed Riemann–Liouville (RL) fractional derivative* (see [13]). In this fractional derivative α is its real-number order.

5.2 *Two Transformation Rules from Locality to Nonlocality*

For quasi-steady turbulent shear flows, in analogy to the time dependent version (15), a spatially dependent expression of the transformation equation is applied to $\bar{u}_1(x_1, x_2)$

$$\frac{d}{dx_2}\bigg|_{nonlocal} \bar{u}_1(x_1, x_2) = \frac{1}{\Gamma(1-\alpha)}\frac{d}{dx_2}\bigg| \underbrace{\int_{-\infty}^{x_2}}_{local} (x_2 - \xi)^{-\alpha}\,\bar{u}_1(x_1, \xi)\,d\xi =: {}_L D_+^{\alpha}\bar{u}_1(x_1, x_2), \tag{16a,b}$$

where, by a comparison of (16a) with (15), the analogue time-space quantities should become clear. Furthermore, the definition of the occurring RL fractional derivative is defined by (16b).

The same process applied to a Heaviside distribution as weighting 'function' (Table 1) leads to

$$\frac{d}{dx_2}\bigg|_{nonlocal} \bar{u}_1(x_1, x_2) = \frac{\int_{-\infty}^{\infty}\theta\left(\lambda, x_2 - x_2'\right)\frac{\partial\bar{u}_1(x_1,x_2')}{\partial x_2'}dx_2'}{\int_{-\infty}^{\infty}\theta\left(\lambda, x_2 - x_2'\right)dx_2'} = \frac{\int_{x_2}^{x_{2\,\max}}\frac{\partial\bar{u}_1}{\partial x_2'}\left(x_1, x_2'\right)dx_2'}{\int_{x_2}^{x_{2\,\max}}dx_2'} \tag{17a–c}$$

$$= \frac{\bar{u}_{1_{\max}}(x_1) - \bar{u}_1(x_1, x_2)}{x_{2\,\max} - x_2}.$$

The carrier domains of the Heaviside distributions in the eddy diffusivity and the up-gradient driving term are discussed in [15]; the second leads to (17c). If in a turbulence model the mixing length χ_2 is e.g. $\chi_2 = (x_{2\,\max} - x_2)$, then the difference quotient simplifies to a mean velocity difference.

5.3 Four New Nonlocal and Fractional Turbulence Models

Now the above two generalized operators are systematically applied to create new nonlocal turbulence models. To this end, the derivatives in the local turbulence models are replaced by (16b), (17c) or its simplification, respectively. Then *Prandtl's Nonlocal Mixing-Length Turbulence Model* is:

Model 1 $$\overline{u_2'u_1'}(x_1, x_2) = -\sigma\,\chi_2^2\left|{}_L D_+^{\alpha}\bar{u}_1(x_1, x_2)\right|{}_L D_+^{\alpha}\bar{u}_1(x_1, x_2). \tag{18}$$

Furthermore, *Prandtl's Modified Nonlocal Shear-Layer Turbulence Model* takes the form:

Model 2 $$\overline{u_2'u_1'}(x_1, x_2) = -\sigma\,\chi_2\left[\bar{u}_1(x_1, x_2) - \bar{u}_{1\,\min}(x_1)\right]_L D_+^{\alpha}\bar{u}_1(x_1, x_2). \tag{19}$$

Note that it looks as if Prandtl already would have applied our second fractional calculus transformation method (17c) in his eddy diffusivity development from 1925 (5a) to 1942 (5b)! However, this was hardly the case; moreover, it demonstrates an extremely high intuitive strength of this leading scientist of aerodynamics. A third possibility would be created by a local zero-equation turbulence model that at early times (1920s), was not proposed, but follows in a natural manner from a systematic consideration

$$\textbf{\textit{Model 3}} \quad \overline{u_2'u_1'}(x_1, x_2) = -\sigma \chi_2^2 {}_L D_+^{\alpha} \bar{u}_1(x_1, x_2) \frac{\bar{u}_{1\max}(x_1) - \bar{u}_1(x_1, x_2)}{x_{2\max} - x_2}. \tag{20}$$

The fourth and last proposed model is the *Difference-Quotient Turbulence Model*

$$\textbf{\textit{Model 4}} \quad \overline{u_2'u_1'}(x_1, x_2) - \sigma \chi_2 \left[\bar{u}_1(x_1, x_2) - \bar{u}_{1\min}(x_1)\right] \frac{\bar{u}_{1\max}(x_1) - \bar{u}_1(x_1, x_2)}{x_{2\max} - x_2}. \tag{21}$$

This closure was discussed by Egolf and Weiss (e.g. see [15]). They wrote: *"Today, in turbulence research, we have some conceptual understanding of such an approach* [Eq. (17a)], *even though a detailed mathematical derivation from basic equations is, to our knowledge, yet to be found.* In our opinion a modeling of turbulence by Lévy walks and the fractal-β model [16], together with the present article, give a very satisfying explanation for the application of convolution integral models as they were proposed e.g. in non-Newtonian fluid dynamics (rheology) and turbulence research; at that time being rather empirical. However, the application of two different weighting functions to develop a nonlocal Reynolds shear stress is not very realistic. Therefore, the first and fourth presented model must ought to be favored. With the argument of Prandtl's intuitive development of a mean-velocity-difference eddy diffusivity, finally a preference must be given to the DQTM. In [17]) for four elementary turbulent shear flows in 12 figures excellent agreement is demonstrated between model results, based on the DQTM, and measurements. This model also reveals a critical phenomenon [16–18].

References

1. R. Dedekind, M. Weber (ed.), Bernard Riemann's gesammelte mathematische Werke und wissenschaftlicher Nachlass, Chap. XIX: Versuch einer allgemeinen Auffassung der Integration und Differentiation. Cambridge University Press, Cambridge (2013)
2. R.K. Raina, C.L. Koul, On Weyl fractional calculus. Proc. Am. Math. Soc. **73**(2), 188–192 (1979)
3. B.B. Mandelbrot, *The Fractal Geometry of Nature* (Freeman and Co., New York, 1977)
4. C. Beck, F. Schlögl, *Thermodynamics of Chaotic Systems*. Cambridge Nonlinear Science Series (Cambridge University Press, New York, 1993)
5. B.J. West, *Fractional Calculus—View of Complexity: Tomorrow's Science* (CRC Press, Taylor & Francis Group, Boca Raton, 2016)
6. P.W. Egolf, K. Hutter, From linear and local to nonlinear and nonlocal zero-equation turbulence models, in *Proceedings of IMA Conference for Turbulence, Waves and Mixing* (Kings College, Cambridge, 2016), pp. 71–74, 6–8 July 2016

7. O. Reynolds, On the dynamical theory of incompressible viscous fluids and the determination of the criterion. Phil. Trans. R. Soc. Lond. A **186**, 123–164 (1895)
8. K. Hutter, K. Jöhnk, *Continuum Methods of Physical Modeling* (Springer, Berlin, 2004)
9. J.O. Hinze, *Turbulence*, 2nd edn (McGraw-Hill, New York, 1975)
10. R.H. Kraichnan, The Closure Problem of Turbulent Theory. Research Report of Office of Naval Research No. HSN-3, pp. 1-48 (1961)
11. L. Prandtl, Bericht über Untersuchungen zur ausgebildeten Turbulenz. ZAMM **5**(2), 136–139 (1925)
12. L. Prandtl, Bemerkungen zur Theorie der freien Turbulenz. ZAMM **22**(5), 241–243 (1942)
13. R. Herrmann, *Fractional Calculus* (World Scientific, Singapore, 2011)
14. I.M. Gelfand, G.E. Shilov, *Generalized Functions*, vol. 1 (American Mathematical Society, Providence, 1964)
15. P.W. Egolf, D.A. Weiss, Difference-quotient turbulence model: the axi-symmetric isothermal jet. Phys. Rev. E **58**(1), 459–470 (1998)
16. P.W. Egolf, Lévy statistics and beta model: a new solution of "wall" turbulence with a critical phenomenon. Int. J. Refrig. **32**, 1815–1836 (2009)
17. K. Hutter, Y. Wang, *Fluid and Thermodynamics, volume 2: Advanced Fluid Mechanics and Thermodynamic Fundamentals*, Chap. 16 (Springer, Berlin, 2016)
18. P.W. Egolf, K. Hutter, in *Turbulent Shear Flow Described by the Algebraic Difference-Quotient Turbulence Model*, ed, by J. Peinke, G. Kampers, M. Oberlack, M. Waclawczyk, A. Talamelli. Progress in Turbulence VI, Springer Proceedings in Physics (Springer, Heidelberg, 2016), pp. 105–109

A Framework for the Assessment and Creation of Subgrid-Scale Models for Large-Eddy Simulation

Maurits H. Silvis, Ronald A. Remmerswaal and Roel Verstappen

Abstract We focus on subgrid-scale modeling for large-eddy simulation of incompressible turbulent flows. In particular, we follow a systematic approach that is based on the idea that subgrid-scale models should preserve fundamental properties of the Navier–Stokes equations and turbulent stresses. To that end, we discuss the symmetries and conservation laws of the Navier–Stokes equations, as well as the near-wall scaling, realizability and dissipation behavior of the turbulent stresses. Regarding each of these properties as a model constraint, we obtain a framework that can be used to assess existing and create new subgrid-scale models. We show that several commonly used velocity-gradient-based subgrid-scale models do not exhibit all the desired properties. Although this can partly be explained by incompatibilities between model constraints, we believe there is room for improvement in the properties of subgrid-scale models. As an example, we provide a new eddy viscosity model, based on the vortex stretching magnitude, that is successfully tested in large-eddy simulations of turbulent plane-channel flow.

1 Introduction

The Navier–Stokes equations form a very accurate model for fluid flows. This model, however, does not form a tractable model, because in general not enough computational power is available to predict the behavior of practical turbulent flows with it. We therefore focus on large-eddy simulation, which aims at predicting the large-scale behavior of turbulent flows. In large-eddy simulation, the large scales of motion in a flow are explicitly computed, whereas small-scale motions are modeled.

M.H. Silvis (✉) · R.A. Remmerswaal · R. Verstappen
University of Groningen, Nijenborgh 9, 9747 AG Groningen, The Netherlands
e-mail: m.h.silvis@rug.nl

R.A. Remmerswaal
e-mail: r.a.remmerswaal@rug.nl

R. Verstappen
e-mail: r.w.c.p.verstappen@rug.nl

R. Örlü et al. (eds.), *Progress in Turbulence VII*, Springer Proceedings in Physics 196, DOI 10.1007/978-3-319-57934-4_19

In the current work, we address the question of how to construct subgrid-scale models for these small-scale motions in turbulent flows. To answer this question, we follow a systematic approach based on the idea that it is desirable that subgrid-scale models are consistent with the physical and mathematical properties of the Navier–Stokes equations and the turbulent stresses. These properties can therefore be seen as requirements for subgrid-scale modeling and we will use them to assess existing and construct new subgrid-scale models.

The structure of this paper is as follows. In Sect. 2 we describe several properties of the Navier–Stokes equations and turbulent stresses, and we discuss their importance. This leads to a framework of model requirements that, in Sect. 3, is used to analyze the properties of existing subgrid-scale models. Finally, in Sect. 4 we give an example of a new eddy viscosity model that can be derived from the model constraints and we test it in large-eddy simulations of turbulent plane-channel flow.

2 Model Constraints for Large–Eddy Simulation

In large-eddy simulation, the large-scale behavior of incompressible turbulent flows is described by the filtered incompressible Navier–Stokes equations [15],

$$\frac{\partial \bar{u}_i}{\partial t} + \frac{\partial}{\partial x_j}(\bar{u}_i \bar{u}_j) = -\frac{1}{\rho}\frac{\partial \bar{p}}{\partial x_i} + \nu \frac{\partial^2 \bar{u}_i}{\partial x_j \partial x_j} - \frac{\partial}{\partial x_j}\tau_{ij}, \quad \frac{\partial \bar{u}_i}{\partial x_i} = 0. \tag{1}$$

The turbulent stresses, $\tau_{ij} = \overline{u_i u_j} - \bar{u}_i \bar{u}_j$, are not solely expressed in terms of the filtered velocity field and therefore have to be modeled. In what follows we will discuss a number of fundamental properties of the Navier–Stokes equations and the turbulent stresses that lead to constraints for this modeling process. More detailed information about these properties and the resulting constraints for subgrid-scale models can be found in previous work [16].

Symmetries of the Incompressible Navier–Stokes Equations The incompressible Navier–Stokes equations are form invariant under several coordinate transformations [11, 12]. Such transformations, or symmetries, play an important role because they ensure that the description of fluids is the same in all inertial frames of reference. They also relate to conservation and scaling laws [13]. Speziale [18], Oberlack [11, 12] and Razafindralandy et al. [13] therefore argue that it is desirable that these symmetries are preserved by subgrid-scale models. We distinguish invariance under the time (S1) and pressure (S2) translations, the generalized Galilean transformation (S3), rotations and reflections (S4), scaling transformations (S5), two-dimensional material frame-indifference (S6) and time reversal (S7) [11, 12].

Conservation Laws Even though the incompressible Navier–Stokes equations are inherently dissipative, they obey several conservation laws. In particular, we have conservation of generalized linear momentum (C1), conservation of angular momentum (C2) and conservation of an infinite hierarchy of vorticity-related quantities (C3) [3]. Conservation laws should not be violated by subgrid-scale models.

Near-Wall Scaling of the Turbulent Stresses Using numerical simulations, Chapman and Kuhn [2] have revealed the near-wall scaling of the time-averaged turbulent stresses. Focusing on wall-resolved large-eddy simulations, we would like to make sure that modeled stresses exhibit the same near-wall scaling behavior (N). In particular, the desired scaling of an eddy viscosity is $\nu_\mathrm{e} = \mathcal{O}(x_2^3)$, where x_2 represents the wall-normal coordinate. This ensures that dissipative effects due to the model fall off quickly enough near solid boundaries.

Realizability of the Turbulent Stresses Vreman et al. [23] showed that, for positive spatial filters, the turbulent stress tensor, τ_{ij}, is realizable, i.e., it has no negative eigenvalues. As these eigenvalues can be interpreted as (partial) energies, it seems desirable that subgrid-scale models exhibit realizability (R) as well.

Production of Subgrid-Scale Kinetic Energy Subgrid-scale models generally increase the dissipation of large-scale kinetic energy, i.e., the transport of energy from large to small scales of motion. We now focus on this process, which is also referred to as the production of subgrid-scale kinetic energy.

Vreman's Requirements Vreman [22] showed that the production of subgrid-scale kinetic energy due to the true turbulent stresses is zero for certain (laminar) flows. He therefore argues that the production due to subgrid-scale models should also be zero for these flows (P1a). On the other hand, subgrid-scale models should not turn off in regions of flows where turbulence occurs (P1b). This ensures that subgrid-scale models are neither overly nor underly dissipative.

Nicoud et al. Requirements On the basis of physical grounds, Nicoud et al. [10] reason that certain flows cannot be maintained if energy is transported to subgrid scales. They therefore see it as a desirable property that the modeled production of subgrid-scale kinetic energy vanishes for these flows. In particular, they require that a model's production of subgrid-scale kinetic energy vanishes for all two-component flows (P2a) and for the pure axisymmetric strain (P2b). Note that requirement P2a is not compatible with P1b, because the latter requires that certain two-component flows have a nonzero production of subgrid-scale kinetic energy [16].

The Second Law of Thermodynamics In turbulent flows, energy can be transported from large to small scales (forward scatter) and vice versa (backscatter). The second law of thermodynamics requires that the net transport is of the former type (P3) [13].

Verstappen's Requirements Verstappen [20] argues that large-eddy simulation is ultimately aimed at predicting large-scale flow dynamics, independent of small-scale motions. Therefore, subgrid-scale models have to cause scale separation. This can be achieved by ensuring that subgrid-scale models are sufficiently dissipative, such that they counterbalance the convective production of small-scale kinetic energy and dissipate any kinetic energy (initially) contained in small scales of motion (P4). Requirements P4 and P2b cannot be satisfied at the same time, because the former requires a nonzero dissipation for the axisymmetric strain [16].

3 Analysis of Existing Subgrid-Scale Models

With the list of fundamental properties of Sect. 2, we obtain a framework that can be used to assess the behavior of subgrid-scale models. Table 1 provides a summary of the analysis of some commonly used velocity-gradient-based subgrid-scale models.

Velocity-gradient-based subgrid-scale models automatically preserve certain symmetries (S1–S4). Scaling invariance (S5), however, is usually violated because of the use of the local grid size as characteristic length scale [11, 13]. The dynamic procedure [5] may restore scaling invariance [1, 11, 13]. The importance of two-dimensional material frame-indifference (S6) is disputed [12], while time reversal invariance (S7) is generally not regarded as a desirable property of subgrid-scale models [1]. The three conservation laws (C1–C3) are trivially preserved for symmetric subgrid-scale models appearing in the form $\partial/\partial x_j\ \tau_{ij}^{\mathrm{mod}}$. Realizability (R) does not pertain to traceless subgrid-scale models, including the eddy viscosity models studied here.

Table 1 Summary of the properties of several subgrid-scale models. The properties considered are S1–4: time, pressure, generalized Galilean, and rotation and reflection invariance; S5: scaling invariance; S6: two-dimensional material frame-indifference; S7: time reversal invariance; C1: conservation of generalized linear momentum; C2: conservation of angular momentum; C3: conservation of vorticity-related quantities; N: the proper near-wall scaling behavior; R: realizability; P1a: zero subgrid dissipation for laminar flow types; P1b: nonzero subgrid dissipation for nonlaminar flow types; P2a: zero subgrid dissipation for two-component flows; P2b: zero subgrid dissipation for the pure axisymmetric strain; P3: consistency with the second law of thermodynamics; P4: sufficient subgrid dissipation for scale separation

	S1–4	S5[a]	S6	S7[a]	C1	C2	C3	N[a]	R	P1a	P1b	P2a	P2b	P3	P4
Smagorinsky [17]	Y	N	Y	N	Y	Y	Y	N		N	Y	N	N	Y	Y
WALE [9]	Y	N	N	N	Y	Y	Y	Y		N	Y	N	N	Y	Y
Vreman [22]	Y	N	N	N	Y	Y	Y	N		N	Y	N	N	Y	Y
σ [10]	Y	N	Y	N	Y	Y	Y	Y		Y	N	Y	Y	Y	N
QR [20]	Y	N	Y	N	Y	Y	Y	N		Y	N	Y	N	Y	N
S3PQR [19]	Y	N	Y[b]	N[b]	Y	Y	Y	Y		Y[b]	Y[b]	Y[b]	N	Y[b]	Y[b]
AMD [14]	Y	N	Y	N	Y	Y	Y	N		Y	N	Y	N	Y	Y
Vortex stretching	Y	N	Y	N	Y	Y	Y	Y		Y	N	Y	Y	Y	N
Gradient [4, 6]	Y	N	N	Y	Y	Y	Y	N	Y	Y	N	Y	N	N	
EASSM [7]	Y	N	N	N	Y	Y	Y	N	Y	N	Y	N	N	Y	

[a] The dynamic procedure [5] may restore these properties [1, 11, 13]
[b] Depending on the value of the model parameter and/or the implementation

The general view that we obtain from Table 1 is that existing subgrid-scale models do not satisfy all the desired properties. This can partly be understood from incompatibilities between model constraints, especially the different dissipation requirements, and from difficulties with satisfying scale invariance. We do, however, believe that there is room for improvement in the properties of subgrid-scale models that are based on the velocity gradient.

4 Example of a New Subgrid-Scale Model

The framework of model constraints can also be used to create new subgrid-scale models. For example, we previously derived the vortex-stretching-based (VS) eddy viscosity model [16],

$$\tau_{ij}^{\mathrm{mod,dev}} = -2(C_{\mathrm{VS}}\delta)^2\sqrt{2\mathrm{tr}(\bar{S}^2)}\left(\frac{\mathrm{tr}(\bar{S}^2\bar{\Omega}^2) - \frac{1}{2}\mathrm{tr}(\bar{S}^2)\mathrm{tr}(\bar{\Omega}^2)}{-\mathrm{tr}(\bar{S}^2)\mathrm{tr}(\bar{\Omega}^2)}\right)^{3/2} \bar{S}_{ij} \,. \quad (2)$$

Here, C_{VS} is a model constant, whereas δ denotes the characteristic length scale of the large-eddy simulation. $\bar{S}$ and $\bar{\Omega}$ represent the rate-of-strain and rate-of-rotation tensors, i.e., the symmetric and asymmetric parts of the velocity gradient, $\partial\bar{u}_i/\partial x_j$. The quantity $4(\mathrm{tr}(\bar{S}^2\bar{\Omega}^2) - \frac{1}{2}\mathrm{tr}(\bar{S}^2)\mathrm{tr}(\bar{\Omega}^2))$ is the (squared) vortex stretching magnitude [19], which corrects for the dissipation behavior and the near-wall scaling of the Smagorinsky model.

Figure 1 shows results of large-eddy simulations of turbulent plane-channel flow obtained using the vortex-stretching-based eddy viscosity model. These simulations were performed using an incompressible Navier–Stokes solver that employs a symmetry-preserving finite-volume discretization on a staggered grid [21]. A 64^3 grid was used that was stretched in the wall-normal direction. The value of the model constant, $C_{\mathrm{VS}} \approx 0.58$, was obtained by matching the average model dissipation with that of the Smagorinsky model [10, 19]. The mean velocity in the near-wall region is predicted remarkably well for this C_{VS}. Also the location of the peaks in the Reynolds stresses and the behavior of the stresses in the center of the channel is predicted well. The underpredicted center line velocity and the over- and undershoots in the Reynolds stresses seem to be common deficiencies of eddy viscosity models. All in all, these encouraging results show how new subgrid-scale models with built-in desirable properties can be constructed.

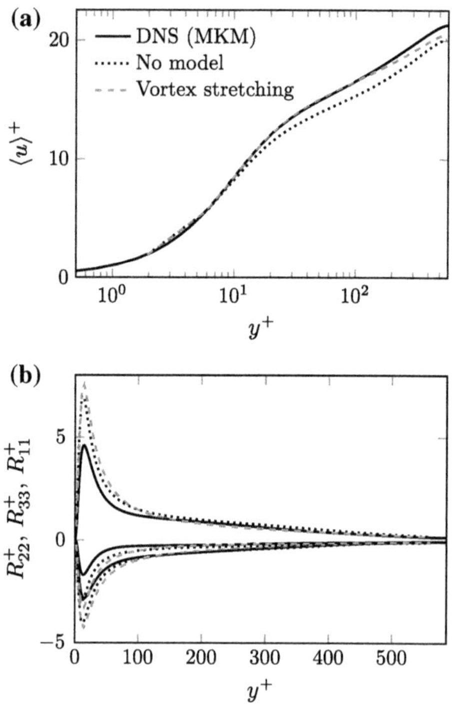

Fig. 1 **a** Mean velocity profile and **b** diagonal deviatoric Reynolds stresses compensated by the average model contribution, as obtained from large-eddy simulations of turbulent plane-channel flow at $Re_\tau \approx 590$ on a 64^3 grid. Simulations were performed without a subgrid-scale model (*dotted line*) and with the vortex-stretching-based eddy viscosity model (*dashed line*) of (2) with $C_{VS} \approx 0.58$. Results from direct numerical simulations (DNS) [8] are shown as reference (*solid line*). All quantities are shown in wall units

Acknowledgements The authors thankfully acknowledge Professor Martin Oberlack for stimulating discussions during several stages of this project. Professor Michel Deville is kindly acknowledged for sharing his insights relating to nonlinear subgrid-scale models and realizability. This work is part of the research programme Free Competition in the Physical Sciences with project number 613.001.212, which is financed by the Netherlands Organisation for Scientific Research (NWO).

References

1. D. Carati, G.S. Winckelmans, H. Jeanmart, J. Fluid Mech. **441**, 119 (2001)
2. D.R. Chapman, G.D. Kuhn, J. Fluid Mech. **170**, 265 (1986)
3. A.F. Cheviakov, M. Oberlack, J. Fluid Mech. **760**, 368 (2014)
4. R.A. Clark, J.H. Ferziger, W.C. Reynolds, J. Fluid Mech. **91**, 1 (1979)
5. M. Germano, U. Piomelli, P. Moin, W.H. Cabot, Phys. Fluids A **3**, 1760 (1991)
6. A. Leonard, in *Turbulent Diffusion*, ed. by F.N. Frenkiel, R.E. Munn. Environmental Pollution, Adv. Geophys., vol. 18 A (Academic Press, New York, 1974), pp. 237–248
7. L. Marstorp, G. Brethouwer, O. Grundestam, A.V. Johansson, J. Fluid Mech. **639**, 403 (2009)
8. R.D. Moser, J. Kim, N.N. Mansour, Phys. Fluids **11**, 943 (1999)
9. F. Nicoud, F. Ducros, Flow Turbul. Combust. **62**, 183 (1999)
10. F. Nicoud, H. Baya Toda, O. Cabrit, S. Bose, J. Lee, Phys. Fluids **23**, 085106 (2011)
11. M. Oberlack, *Annual Research Briefs* (Center for Turbulence Research, Stanford University, Stanford, 1997)

12. M. Oberlack, in *Theories of Turbulence*, ed. by M. Oberlack, F. Busse. International Centre for Mechanical Sciences, vol. 442 (Springer, Vienna, 2002), pp. 301–366
13. D. Razafindralandy, A. Hamdouni, M. Oberlack, Eur. J. Mech. B **26**, 531 (2007)
14. W. Rozema, H.J. Bae, P. Moin, R. Verstappen, Phys. Fluids **27**, 085107 (2015)
15. P. Sagaut, *Large Eddy Simulation for Incompressible Flows*, 3rd edn. (Springer, Berlin, 2006)
16. M.H. Silvis, R.A. Remmerswaal, R. Verstappen, Phys. Fluids **29**, 015105 (2017)
17. J. Smagorinsky, Mon. Weather Rev. **91**, 99 (1963)
18. C.G. Speziale, J. Fluid Mech. **156**, 55 (1985)
19. F.X. Trias, D. Folch, A. Gorobets, A. Oliva, Phys. Fluids **27**, 065103 (2015)
20. R. Verstappen, J. Sci. Comput. **49**, 94 (2011)
21. R.W.C.P. Verstappen, A.E.P. Veldman, J. Comput. Phys. **187**, 343 (2003)
22. A.W. Vreman, Phys. Fluids **16**, 3670 (2004)
23. B. Vreman, B. Geurts, H. Kuerten, J. Fluid Mech. **278**, 351 (1994)

Towards an Improved Subgrid-Scale Model for Thermally Driven Flows

Riccardo Togni, Andrea Cimarelli and Elisabetta De Angelis

Abstract The effect of the filtering on the resolved and subgrid dynamics of turbulent Rayleigh–Bénard convection (RBC) is studied a priori using a Direct Numerical Simulation dataset. To this end, the velocity and temperature fields, split into resolved and subgrid components by a spectral cutoff filter, are analyzed with the filtered turbulent kinetic energy and temperature variance budgets. At small filter lengths, the resolved processes correspond to the exact ones except for the decreases of the dissipations which, in turn, are balanced by the sink actions of the subgrid scales. At large filters lengths, the resolved dynamics depletes close to the walls and the effect of the subgrid scales drifts from purely-dissipative to a more complex behaviour. This study highlights the possibility that eddy-viscosity and diffusivity models, commonly employed in large-eddy simulation of RBC, does not work well for large filter widths and that alternative closures should be considered.

1 Introduction

Turbulent flows relevant in Nature and technology are still out of reach of Direct Numerical Simulations (DNS) due to the immense number of degrees of freedom that need to be solved. However, the steady increase of computational power during the last years gave a boost to a different approach, namely the Large Eddy Simulation (LES), which represents a good compromise between DNS and the solution of the Reynolds-averaged Navier–Stokes (RANS) in terms of accuracy and cost. The basic philosophy of LES is to compute the large, energy-carrying scales and to model the

R. Togni
DIN, Università di Bologna, Via Fontanelle 40, 47121 Forlì, Italy

A. Cimarelli
DISMI, Università degli Studi di Modena e Reggio Emilia, Via Giovanni Amendola 2, 42122 Reggio Emilia, Italy

E. De Angelis (✉)
School of Engineering, Cardiff University, Queen's Buildings, The Parade, Cardiff CF24 3AA, UK
e-mail: DeAngelisE@cardiff.ac.uk

R. Örlü et al. (eds.), *Progress in Turbulence VII*, Springer Proceedings in Physics 196, DOI 10.1007/978-3-319-57934-4_20

effect of the small, more universal ones. For this purpose, the governing equations are filtered, the unclosed terms arising from non-linearity are parameterized using filtered quantities and the system is solved numerically [1].

The filter length separates the large, resolved scales from the small, subgrid ones, thus it represents a crucial parameter in LES as it determines which dynamics is going to be captured and which one should be accounted by the subgrid-scale (SGS) model. The main goal of this work is to investigate a priori the effect of the filter length on the resolved and subgrid dynamics of turbulent Rayleigh–Bénard convection (RBC). The system is a classic paradigm of natural convection and it consists of a fluid layer between two infinite plates, heated from below and cooled from above [2]. Following the classic a priori approach, the velocity and temperature fields obtained from DNS are split into resolved and subgrid components by applying a spectral cutoff filter along the wall-parallel directions [3]. The resulting dataset is analyzed via filtered turbulent kinetic energy and temperature variance budgets to assess the two distinct, but closely related, effects of the filter length on LES results, namely the capability to capture the real dynamics and the role of the subgrid scales.

We refer to a DNS dataset of turbulent RBC obtained by solving the Boussinesq equations with a pseudo-spectral method. The momentum, continuity and temperature budgets are non-dimensionalized using the height of the fluid layer H, the temperature difference between the lower and the upper plates $\Delta\Theta$ and the free-fall velocity $U_f = \sqrt{g\alpha\Delta\Theta H}$, where g is the gravitational acceleration and α is the thermal expansion coefficient. The computational domain is a rectangular cell of sizes $8 \times 8 \times 1$ along the wall-parallel directions x, y and the wall-normal one z, where the Cartesian coordinate system is centered at the lower plate with the z-axis pointing in the direction opposite to the gravity acceleration. Periodic boundary conditions are imposed at the sidewalls whereas isothermal and no-slip boundary conditions are used on the horizontal plates. The DNS is conducted at a moderate Rayleigh number, $Ra = 1.0 \times 10^7$, and for Prandtl number (Pr) equal to 0.7 [4]. Three filter lengths are considered in this study, $\ell_F^* = 7.7$, 16.7 and 25.5, where the asterisk denotes the normalization with respect to the volume-averaged Kolmogorov length scale. It is also worth remarking that the smallest filter length corresponds to two-times the thermal boundary layer thickness δ_θ based on the slope of the average temperature profile Θ, i.e. $\delta_\theta = 0.5/(d\Theta/dz)|_{z=0}$.

2 Analysis of Filtered Budgets

The budgets of filtered turbulent kinetic energy $\langle\overline{k}\rangle = 0.5\langle\overline{u}_i\overline{u}_i\rangle$ and filtered temperature variance $\langle\overline{\theta}^2\rangle$ are in order

$$-\frac{d\left\langle\overline{k}\overline{w}\right\rangle}{dz} - \frac{d\left\langle\bar{w}\bar{p}\right\rangle}{dz} + \sqrt{\frac{Pr}{Ra}}\frac{d^2\left\langle\overline{k}\right\rangle}{dz^2} + \left\langle\overline{w}\overline{\theta}\right\rangle - \left\langle\overline{\varepsilon}\right\rangle - \frac{d\left\langle\tau_{i3}\overline{u}_i\right\rangle}{dz} - \left\langle\varepsilon_{sgs}\right\rangle = 0, \quad (1)$$

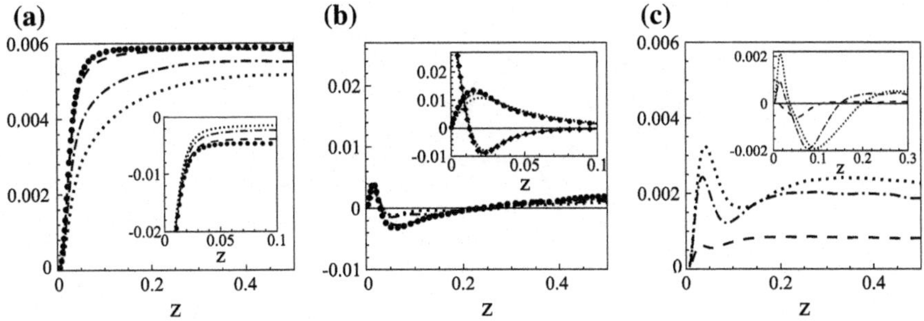

Fig. 1 **a** (*Main panel*) production and (*inset*) dissipation of unfiltered and filtered turbulent kinetic energy versus z. **b** (*Main panel*) inertial, (*inset*) pressure and viscous transports of unfiltered and filtered turbulent kinetic energy versus z. **c** (*Main panel*) average subgrid dissipation and (*inset*) SGS transport versus z. *Different lines* and *symbols* correspond to: unfiltered quantities (*symbols*), $\ell_F^* = 7.7$ (*dashed line*), $\ell_F^* = 16.7$ (*dot-dashed line*) and $\ell_F^* = 25.5$ (*dotted line*). In the *inset* of (**b**), the unfiltered pressure and viscous transports are denoted with *circles* and *diamonds* respectively

$$-\frac{d\left\langle\overline{\theta}^2\overline{w}\right\rangle}{dz}+\frac{1}{\sqrt{Pr\,Ra}}\frac{d^2\left\langle\overline{\theta}^2\right\rangle}{dz^2}-2\left\langle\overline{w\theta}\right\rangle\frac{d\Theta}{dz}-\langle\overline{\chi}\rangle-2\frac{d\left\langle q_3\overline{\theta}\right\rangle}{dz}-\left\langle\chi_{sgs}\right\rangle=0, \quad (2)$$

where $\langle\cdot\rangle$ denotes the ensemble average, $\overline{u}_i$, $\overline{p}$, $\overline{\theta}$ are the filtered (or resolved) fluctuating velocity, pressure and temperature fields, $\overline{\varepsilon}=\sqrt{Pr/Ra}\left(\partial\overline{u}_i/\partial x_j\right)\left(\partial\overline{u}_i/\partial x_j\right)$ and $\overline{\chi}=2\left[\left(\partial\overline{\theta}/\partial x_i\right)\left(\partial\overline{\theta}/\partial x_i\right)\right]/\sqrt{Pr\,Ra}$ are respectively the pseudo-dissipation of filtered turbulent kinetic energy and the dissipation of filtered temperature variance, $\varepsilon_{sgs}=-\tau_{ij}\overline{S}_{ij}$ and $\chi_{sgs}=-2q_i\overline{Q}_i$ are the subgrid dissipations, $\overline{S}_{ij}$ is the resolved strain-rate tensor, $\overline{Q}_i$ is the resolved gradient of θ, $\tau_{ij}=\overline{u_iu_j}-\overline{u}_i\overline{u}_j$ is the SGS stress tensor and $q_i=\overline{u_i\theta}-\overline{u}_i\overline{\theta}$ is the SGS heat flux.

The terms of the equations can be categorized into sources/sinks (viz. the productions $\langle\overline{w\theta}\rangle$ and $-2\langle\overline{w\theta}\rangle d\Theta/dz$, the average dissipations $\langle\overline{\varepsilon}\rangle$ and $\langle\overline{\chi}\rangle$, the average subgrid dissipations $\langle\varepsilon_{sgs}\rangle$ and $\langle\chi_{sgs}\rangle$) and transports along z (viz. inertial $-d\langle\overline{k}\overline{w}\rangle/dz$, $-d\langle\overline{\theta}\overline{w}\rangle/dz$, viscous $(\sqrt{Pr/Ra})d^2\langle\overline{k}\rangle/dz^2$, $(1/\sqrt{Pr\,Ra})d^2\langle\theta^2\rangle/dz^2$, pressure $-d\langle\bar{w}\bar{p}\rangle/dz$ and SGS transports $-d\langle\tau_{i3}\overline{u}_i\rangle/dz$, $-2d\langle q_3\overline{\theta}\rangle/dz$).

As can be seen in the main plot of Figs. 1a and 2a, the depletion of the resolved productions with respect to the unfiltered counterparts is negligible for $\ell_F^*=7.7$, it increases monotonically with ℓ_F^* and it is always peaked in the near-wall region around $z=0.05$. The insets in Figs. 1a and 2a reveal that the resolved average dissipations are very small in the core of the flow ($z\gtrsim 0.05$) and almost independent from the filter length in the near-wall region ($z\lesssim 0.02$) for large ℓ_F^*, which means that dissipations occur entirely at subgrid level away from the boundary whereas the resolved scales are the most dissipative ones next to the wall.

As well as resolved productions and dissipations, resolved transports overlap the unfiltered counterparts when $\ell_F^*=7.7$, see Figs. 1b and 2b. The component which seems to be more affected by the increase of ℓ_F^* is the inertial transport, whereas viscous and pressure contributions barely deplete with respect to the unfiltered refer-

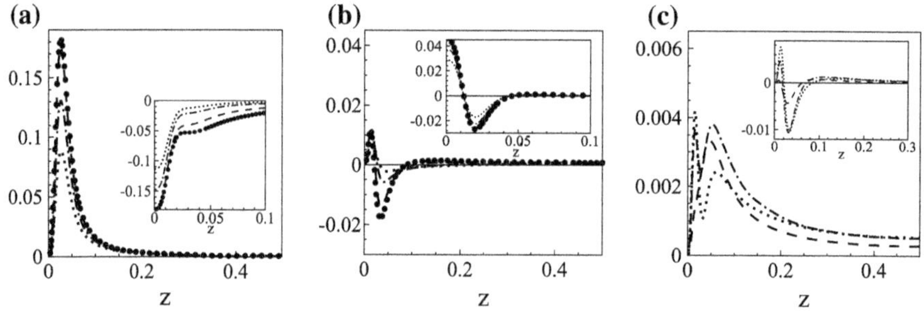

Fig. 2 **a** (*Main panel*) production and (*inset*) dissipation of unfiltered and filtered temperature variance versus z. **b** (*Main panel*) inertial and (*inset*) viscous transports of unfiltered and filtered temperature variance versus z. **c** (*Main panel*) average subgrid dissipation and (*inset*) SGS transport versus z. *Different lines and symbols* correspond to: unfiltered quantities (*symbols*), $\ell_F^* = 7.7$ (*dashed line*), $\ell_F^* = 16.7$ (*dot-dashed line*) and $\ell_F^* = 25.5$ (*dotted line*)

ences. More in details, the energy and temperature variance convected from the region around $z = 0.05$ ($-d\left\langle \overline{kw} \right\rangle/dz < 0$ and $-d\left\langle \overline{\theta w} \right\rangle/dz < 0$) to sustain the core and the near-wall region ($-d\left\langle \overline{kw} \right\rangle/dz > 0$ and $-d\left\langle \overline{\theta w} \right\rangle/dz > 0$) is especially subjected to depletion. This phenomenon is possibly related to the difficulty encountered by wide filters in capturing the dynamics of the coherent structures which populate RBC, the so-called the thermal plumes. Indeed, these structures have a diameter comparable to the thermal boundary layer thickness and they carry most of the turbulent fluctuation throughout the domain [5].

The average subgrid dissipations $\langle \varepsilon_{sgs} \rangle$ and $\langle \chi_{sgs} \rangle$, represented in the main plots of Figs. 1c and 2c, are non-negative at every z and for all ℓ_F^* considered in this study, meaning that subgrid scales behave, on average, as sinks of turbulent kinetic energy and temperature variance. The shape of the profiles change quite evidently passing from small to large ℓ_F^* and this alteration is marked in $\langle \chi_{sgs} \rangle$ where a minimum appears around $z = 0.05$. The subgrid transports are negligible in comparison with the resolved ones for $\ell_F^* = 7.7$ whereas they become relevant for $\ell_F^* = 16.7$ and $\ell_F^* = 25.5$. The SGS redistribution and the corresponding inertial transports have similar profiles for $\ell_F^* = 16.7$ and $\ell_F^* = 25.5$, i.e. the subgrid scales, apart from extracting a net amount of turbulent kinetic energy and temperature variance from the resolved ones ($\langle \varepsilon_{sgs} \rangle > 0$, $\langle \chi_{sgs} \rangle > 0$), transport resolved fluctuations from the transitional layer towards the bulk and the near-wall region.

3 Discussion and Final Remarks

The analysis of the filtered budgets reveals that the mechanisms of turbulent RBC, namely production, dissipation and transport of velocity and temperature fluctuations, are perfectly captured by filter lengths which are comparable to $\ell_F^* = 7.7$.

This observation can be reasonably motivated by the important role played by thermal plumes in the self-sustained cycle of turbulence and by the fact that these structures present a characteristic diameter of the order of the thermal boundary layer thickness. This first result establishes somehow a minimum resolution requirement that the LES should satisfy in order to be well-resolved. Furthermore, since the thermal boundary layer thickness is defined as $\delta_\theta = 0.5/(d\Theta/dz)|_{z=0} = 0.5/Nu$ and the Nusselt number Nu can be estimated from Ra and Pr by means of several and accurate scaling laws [6], the minimum resolution is known a priori.

For small filter lengths, i.e. $\ell_F^* < 7.7$, the only role of the subgrid scales is to drain a certain amount of resolved turbulent kinetic energy and temperature variance and to dissipate it. However, as the filter length increases, a non-negligible part of the production and of the transports start to involve also the subgrid scales, as can be seen from the depletion of the resolved terms plotted in Figs. 1a,b and 2a,b. Hence, the subgrid dynamics is no more characterized only by dissipation but also by inhomogeneous processes which modulate the profiles of the subgrid dissipations. In particular, we see from the inspection of the main plot in Fig. 2c that $\langle \chi_{sgs} \rangle$ exhibit a clear minimum for $\ell_F^* > 7.7$ in correspondence of the maximum depletion region ($z \approx 0.05$) due to the combined effect of production and transport occurring at subgrid level. In parallel, the SGS redistribution effects become important when $\ell_F^* > 7.7$ due to the active role of subgrid scales in transporting resolved turbulent kinetic energy and temperature variance.

In conclusion, while classic eddy-viscosity and diffusivity models are arguably successful in reproducing the energy and temperature variance exchange between large, inhomogeneous scales and small, purely-dissipative ones, they probably fail in case of very large filter due to the complex dynamics of subgrid scales [7]. Further studies will be devoted to the a priori assessment of Smagorsinky-like models for RBC and, eventually, to the development of improved SGS models capable to account for the production and transport mechanisms occurring at subgrid level.

References

1. U. Piomelli, Large-eddy simulation: achievements and challenges. Prog. Aerosp. Sci. **34**(4), 335–412 (1999)
2. F. Chillà, J. Schumacher, New perspectives in turbulent Rayleigh–Bénard convection. Eur. Phys. J. E **35**(7), 1–25 (2012)
3. C. Härtel, L. Kleiser, F. Unger, R. Friedrich, Subgrid-scale energy transfer in the near-wall region of turbulent flows. Phys. Fluids **6**(9), 3130–3130 (1994–present)
4. R. Togni, A. Cimarelli, E. De Angelis, Physical and scale-by-scale analysis of Rayleigh–Bénard convection. J. Fluid Mech. **782**, 380–404 (2015)
5. Q. Zhou, C. Sun, K.Q. Xia, Morphological evolution of thermal plumes in turbulent Rayleigh–Bènard convection. Phys. Rev. Lett. **98**(7), 074501 (2007)
6. S. Grossmann, D. Lohse, Scaling in thermal convection: a unifying theory. J. Fluid Mech. **407**, 27–56 (2000)
7. W.H. Cabot, *Large Eddy simulations of time-dependent and buoyancy-driven channel flows. Annual research briefs* (NASA Ames/Stanford University, Stanford, 1993)

Integral Formula for Determination of the Reynolds Stress in Canonical Flow Geometries

Tae-Woo Lee and Jung Eun Park

Abstract We present a theoretical framework for solving for the Reynolds stress in turbulent flows, based on fundamental physics of turbulence transport. Results thus far indicate that the good agreement between the current theoretical results with experimental and DNS (direct numerical simulation) data is not a fortuitous coincidence, and in the least the current approach is the best hypothesis available in canonical flow geometries. The theory leads to simple and correct expressions for the Reynolds stress in various flow geometries, in terms of the root variables, such as the mean velocity, velocity gradient, turbulence kinetic energy and a viscous term. The applications for this theory are construction of effective turbulence models based on correct physics, and potentially augmenting or replacing turbulence models in simple flows. However, as the method is thus far proven only for relatively simple flow geometries, and implications and nuances for full, three-dimensional flows need to be further examined.

1 Introduction

Turbulence is considered one of the most difficult problems in fluid physics, or some say, physics in general. It is also quite important as many issues of practical concern, such as weather, aerodynamics, combustion flows and many industrial processes depend on turbulence, and much work has been done on finding some adequate approximations so that immediate problems of turbulent flows can be solved (we do not attempt to list the vast literature in this area). As finding the entire absolute (mean + fluctuations) velocity field is quite difficult, or as some argue an overflow of information, here we focus on finding the Reynolds stress as a function of the root turbulence parameters, such as the mean velocity and its gradient, turbulence kinetic energy, in particular its longitudinal component, and also a viscous term.

T.-W. Lee (✉) · J.E. Park
Mechanical and Aerospace Engineering, SEMTE, Arizona State University, Tempe, AZ 85287-6106, USA
e-mail: attwl@asu.edu

R. Örlü et al. (eds.), *Progress in Turbulence VII*, Springer Proceedings in Physics 196, DOI 10.1007/978-3-319-57934-4_21

2 Mathematical Formulation and Results

We start from the Reynolds-averaged Navier–Stokes (RANS) equation. Here, we omit the bar above u'^2, $u'v'$, etc, for simplicity, and take the fluctuation parameters to be time-averaged. Figuring out how the Reynolds stress is related to the mean and other root turbulent parameters has been the topic of numerous studies, for quite some time. However, we notice that the decomposition, $u = U + u'$ that leads to the Reynolds stress term, is necessary only in the absolute coordinate frame. If we move or displace the control volume at the mean speed of the flow (see Fig. 1) then the mean velocity drops out of the momentum equation. That is, RANS is greatly simplified in the relative coordinate frame, or for a control volume moving at the local mean velocity of the fluid. Therefore, the x-momentum equation, for an incompressible boundary-layer flow, becomes:

$$\frac{\partial u'}{\partial t} + \frac{\partial u'^2}{\partial x} + \frac{\partial (u'v')}{\partial y} = -\frac{1}{\rho}\frac{dp'}{dx} + \nu\frac{\partial^2 u'}{\partial y^2} \tag{1}$$

If the time mean of the fluctuating velocity does not vary appreciably in time, then we can write a "steady-state" momentum equation, and solve for the gradient of the Reynolds stress.

$$\frac{\partial (u'v')}{\partial y} = -\frac{\partial u'^2}{\partial x} - \frac{1}{\rho}\frac{dp'}{dx} + \nu\frac{\partial^2 u'}{\partial y^2} \tag{2}$$

In conventional calculations, the x-derivatives would have been set to zero for fully-developed flows, and we would be left with a triviality. However, we note that (1) and (2) have been written for a control volume which is moving along with the mean flow velocity, as shown in Fig. 1, for a boundary-layer flow as an example. For a flow over a flat plate, the boundary layer grows due to the "displacement" effect. The mass is displaced due to the fluid slowing down at the wall, as is the momentum, and it turns out other turbulence parameters as well. The boundary layer thickness grows at a predictable rate, depending on the Reynolds number. Thus, if one rides with the fluid moving at the mean velocity, one would see a change in the all of the turbulence properties, as illustrated in Fig. 1. This displacement effect can be mathematically expressed as:

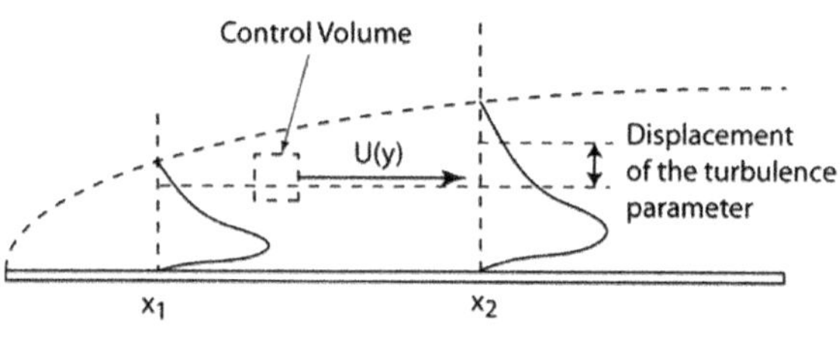

Fig. 1 A schematic illustration of the concept of the theory

$$\frac{\partial}{\partial x} = C_1 U \frac{\partial}{\partial y} \tag{3}$$

I.e., the fluid parcel will see a different portion of the boundary layer in the y-direction, and how much difference it will see depends on how fast the fluid is moving along in the boundary layer. Thus, the mean velocity, U, appears as a multiplicative factor in (3). C_1 is a constant that depends on the Reynolds number. Similarly, the gradient of the pressure fluctuation will not be zero in general. However, this term is expected to be significant only for compressible flows, so we omit this term from further analysis in this phase of the work. In (2), we now have a simple integrable expression to find the Reynolds stress, after using (3). If we integrate by parts, we obtain:

$$u'v' = -C_1 \left[U u'^2 - c \int_0^y \frac{dU}{dy} u'^2 dy \right] + \nu_m \frac{\partial u'}{\partial y} \tag{4}$$

ν_m is the modified kinematic viscosity.

Figure 2 shows a comparison of the Reynolds stress as calculated by (4), with DNS results for fully-developed channel flows [1, 2]. Iwamoto et al. [1, 2] have performed DNS, $Re_\tau = 110$–650, where Re_τ is the Reynolds number based on the friction velocity and channel half-width. The entire data set from the DNS is available on their website [2], including the mean velocity, turbulent fluctuating velocity components, and various moments of their products. We input the necessary root turbulence parameters into (4), and compare with the Reynolds stress from DNS. The agreement is nearly perfect at low Reynolds numbers in Fig. 2, which gives some confidence that we have captured the true physics of turbulent transport, and that the results are not a fortuitous coincidence. The departure at higher Reynolds numbers, as the solution starts to overshoot the DNS data as y approaches the centerline, is due to the displacement effect disappearing near the centerline. Recall that this theory is based on the moving control volume "seeing" a relatively different part of the boundary layer. However, at the centerline there is no displacement effect as all the profiles must be symmetric. One way to impose this kind of symmetry boundary condition is to force the constant C_1 to be proportional to the velocity gradient. For example,

$$C_1 = C_0 \left(\frac{\left(\frac{\partial U}{\partial y}\right)}{\left(\frac{\partial U}{\partial y}\right)_{y=0}} \right)^m \tag{5}$$

Figure 3 shows the comparison of the Reynolds stress obtained from the integral formula (4) with experimental data of DeGraaf and Eaton [4]. In that work, data on various turbulence quantities and the Reynolds stress (all normalized by the friction velocity) are provided, and also various scaling approaches tested with the data, in a well-designed experiment for flows over a flat plate with zero pressure gradient. The Reynolds number based on the momentum thickness (Re_θ) ranged from 1430

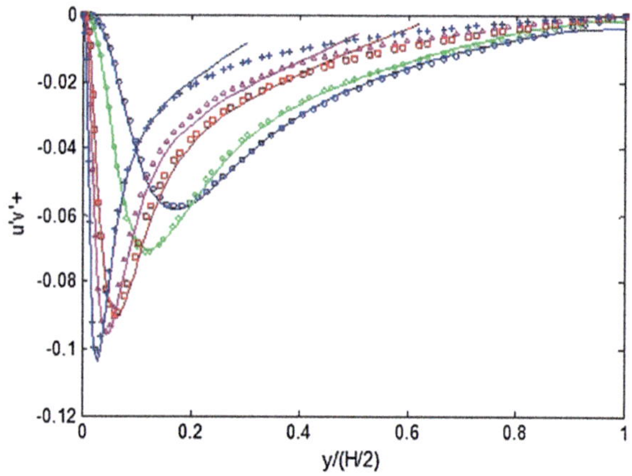

Fig. 2 Comparison of the Reynolds stress in boundary layer flows over a flat plate [3]. *Lines* are theoretical results, using (2). Data symbols: *circle* ($Re_\tau = 110$), *diamond* (150), *square* (300), *triangle* (400) + (650)

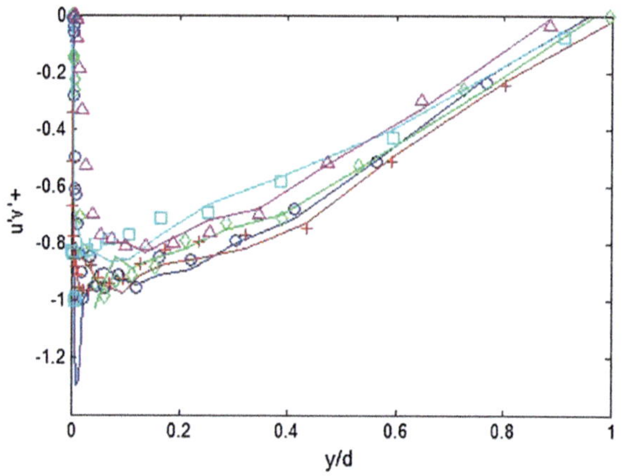

Fig. 3 Comparison of the Reynolds stress in boundary layer flows over a flat plate. The data (*symbols*) are for $Re_\theta = 1430$ ~31000 [4]. *Lines* represent current result (4)

to 31,000 [4]. In (2), we use their measured parameters, U and u'^2, and input them into (2). Gradients of U and u'^2 are calculated from the experimental data. As the experimental data are discontinuous, and at times hard to transcribe, there are some fluctuations and potential errors in the final calculations of the Reynolds stress, particularly close to the wall where the gradients are very steep and the data points all clustered. We can nonetheless input the root parameters into (4) to compute accordingly, and compare with the measured Reynolds stress as in Fig. 3. In spite of dealing with discontinuous experimental data and their gradients, the comparison of (4) result with experimentally observed Reynolds stress is in general quite good.

For axi-symmetric jets, (4) is only slightly modified:

$$u'v' = -C_1\left[Uu_z'^2 - c\int_0^r \frac{dU}{dr}u_z'^2 dy\right] + \nu_m\left(\frac{\partial u_z'}{\partial r} + \int_0^r \frac{1}{r}\frac{\partial u_z'}{\partial r}dr\right) \quad (6)$$

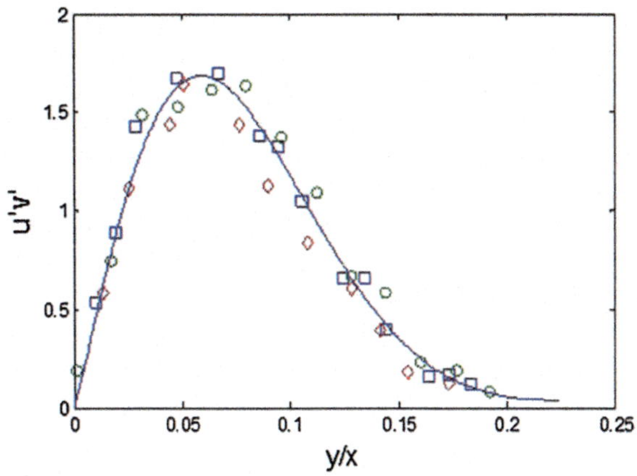

Fig. 4 Reynolds stress in axi-symmetric jets [3]. The experimental data (*symbols*) are from [5]

The only difference from (4) is the second viscous term due to the second derivative in the radial direction. Comparison with experimental data of Wygnanski and Fiedler [5] again produces good agreement in Fig. 4, after applying trajectory correction using (5). This correction is important for jet flows, as the streamlines diverge with increasing z, and therefore the trajectory of the control volume displacement relative to the streamlines must be accounted for through (5). However, this jet divergence is a well-characterized phenomena and $m = 1$ in (5) leads to good agreement with available data as shown in Fig. 4. For internal axi-symmetric flows (pipe flows), the above formula appears to track the Reynolds stress reasonably well near the wall, but undergoes a strong deviation, similar to high Reynolds number cases in Fig. 2. The reason is that the displacement effect is quite strong in pipe flows, as the cumulative effect through the integral term builds radially inward, in comparison to building up from channel wall as in rectangular coordinates.

3 Concluding Remarks

We have shown a theoretical framework for solving for the Reynolds stress in turbulent flows, based on fundamental physics of turbulence transport. Results thus far indicate that the good agreement between the current theoretical and experimental/DNS (direct numerical simulation) data is not a fortuitous coincidence, and that the Reynolds stress is governed by the conservation of momentum principle in the moving coordinate frame, as are all macroscopic objects. The theory leads to simple and correct expressions for the Reynolds stress in various flow geometries, in terms of the root variables, such as the mean velocity, velocity gradient, streamwise turbulence kinetic energy, and a viscous term. However, as the method is thus far proven only for relatively simple, canonical flow geometries, and implications and nuances for full, three-dimensional flows need to be further examined. In that regard, would this method be useful in computing three-dimensional turbulent flows? That is a question that is being thought of at this time. The fact that $u'v'$ is calculated using u'^2 is relatively easy to implement in computational applications as the turbulent kinetic energy can be related to u'^2, assuming isotropy, or an equation for u'^2 can be numerically solved in conjunction with (4). For simple flows, the displacement effect could be treated with (5). Extensions to fully three-dimensional flows will require the displacement effect to be parameterized, which may not be a simple matter. On the other hand, it was considered difficult to parameterize the Reynolds stress even in simple flows, for quite some time.

In summary, (5) illustrates the relatively simple (and perhaps radical, by conventional standards, physics of Reynolds stress production and distribution. The leading term on the RHS of (5) is the dominant term, and is modulated by the viscous term near the wall where again the gradient is extremely high. The integral term reduces the magnitude of the Reynolds stress, particularly at large y locations, where there would otherwise be some deviations. Thus, the turbulent "momentum" or kinetic energy is transported by the mean velocity (the first term on the RHS of (5)), the integral term "displaces" this excess transport term, and the viscous effect dampens the high transport term in region(s) of steep gradients. Therefore, the integral formula contains the "forces and balances" involved in the Reynolds stress production and distribution. Finally, this approach is a unique development in turbulence theory, and is set apart from the Lagrangian statistical approach [6] which is used as data extraction tool for mixing and dispersion properties and for stochastic and pdf modeling of turbulence. The term "integral" is used in data analysis (e.g. [7]); however, this bears no relation to the current method.

References

1. K. Iwamoto, Y. Sasaki, K. Nobuhide, Reynolds number effects on wall turbulence: toward effective feedback control. Int. J. Heat Fluid Flows **23**, 678–689 (2002)
2. K. Iwamoto, N. Kasagi, http://thtlab.jp/DNS/dns_database.html
3. T.-W. Lee, Determination of the Reynolds stress in canonical flows. *APS Fluids Meeting*, Portland, Oregon, November 2016
4. D.B. DeGraaf, J.K. Eaton, Reynolds number scaling of the flat-plate turbulent boundary layer. J. Fluid Mech. **422**, 319–346 (2000)
5. I. Wygnanski, H. Fiedler, Some measurements in the self-preserving jet. J. Fluid Mech. **38**(3), 577–612 (1969)
6. P.K. Yeung, Lagrangian investigation of turbulence. Annu. Rev. Turbul. **34**, 115–142 (2002)
7. F. Mehdi, C.M. White, Integral form of the skin friction coefficient suitable for experimental data. Exp. Fluids **50**(1), 43–51 (2010)

A Matrix-Free Incompressible DG Algorithm for the Simulation of Turbulent Flows

Andrea Crivellini, Matteo Franciolini and Alessandra Nigro

Abstract The paper deals with an efficient implementation of a discontinuous Galerkin solver for the simulation of incompressible turbulent flows. The artificial compressibility flux approach is employed for space discretization, while a Rosenbrock-type Runge–Kutta scheme coupled with a matrix-free linear solver is used for the implicit time integration. The code has been here applied to solve the turbulent Rayleigh–Bénard convection at different Rayleigh numbers. The results prove the reliability of the proposed solution strategy and its suitability for the Implicit Large Eddy Simulation of turbulent flows.

1 Introduction

In recent years discontinuous Galerkin (DG) methods have emerged as one of the most promising high-order discretization techniques for CFD. DG methods have been successfully applied to the simulation of turbulent flows by solving the incompressible and compressible Reynolds Averaged Navier–Stokes (RANS) equations. These methods can easily achieve high-order accuracy on arbitrarily shaped elements and are perfectly suited to hp-adaptation techniques, as well as to parallel computing. Moreover, DG methods show numerical properties suitable for Implicit LES (ILES) of turbulent flows [1, 2]. In fact, the dissipation of the numerical scheme behaves like a spectral cut-off filter, which mimics the role of a subgrid-scale (SGS) model. However, the development of an efficient time integration strategy is still a challenge. On one hand, there exist in literature a variety of algorithms which are implicit and are typically tuned for stiff problems. On the other, the same algorithms may not be

A. Crivellini (✉) · M. Franciolini
Università Politecnica delle Marche, Via Brecce Bianche 12, 60131 Ancona, AN, Italy
e-mail: a.crivellini@univpm.it

M. Franciolini
e-mail: m.franciolini@pm.univpm.it

A. Nigro
Università della Calabria, Ponte P. Bucci Cubo 44/c, 87036 Rende, CS, Italy
e-mail: alessandra.nigro@unical.it

R. Örlü et al. (eds.), *Progress in Turbulence VII*, Springer Proceedings in Physics 196, DOI 10.1007/978-3-319-57934-4_22

enough computationally efficient for the DNS/LES of turbulence. Within this context, the use of the explicit time stepping appears a reliable choice for compressible solvers since the time scales of the turbulent structures are typically small. However, for incompressible solvers, a fully-explicit time stepping strategy cannot be easily adopted since the incompressibility constrain should be fulfilled, and therefore the use of an implicit time integration can still be convenient. Unfortunately, the high CPU-time needed for the evaluation of the Jacobian matrix (it costs up to 290 times a single residual evaluation for a $\mathbb{P}_6$ approximation) and the large amount of memory required for its storage (its size scales approximately as k^6, where k is the order of DG polynomial approximation $\mathbb{P}_k$) hold back the application of this strategy for the simulation of turbulence. To overcome these problems, the present work introduces an approach based on a four stage, order three ROSI2PW Linearly Implicit Rosenbrock Runge–Kutta scheme [3] coupled with a matrix-free GMRES solver [4]. The efficiency of the proposed method and its accuracy have been verified solving the Rayleigh–Bénard convection at Prandtl number equal to 0.7 and different Rayleigh numbers (Ra).

2 The Matrix-Free Algorithm

The Boussinesq governing equations are discretized using the discontinuous Galerkin method described in [5, 6]. Here a matrix-free GMRES solver is employed to avoid the need of the storage of the Jacobian matrix. This algorithm approximates the matrix–vector products of the linear system

$$\left[\frac{\mathbf{M}}{\Delta t} + \frac{\partial \mathbf{R}(\mathbf{U}^n)}{\partial \mathbf{U}}\right] \Delta \mathbf{U} = -\mathbf{R}(\mathbf{U}^n)$$

using a numerical differentiation based on a double evaluation of the residual vector

$$\mathbf{P}\left[\frac{\mathbf{M}}{\Delta t} + \frac{\partial \mathbf{R}(\mathbf{U}^n)}{\partial \mathbf{U}}\right] \Delta \mathbf{U} = \mathbf{P}\frac{\mathbf{M}}{\Delta t}\Delta \mathbf{U} + \mathbf{P}\left(\frac{\mathbf{R}(\mathbf{U}^n + h\Delta \mathbf{U}) - \mathbf{R}(\mathbf{U}^n)}{h}\right),$$

where $\mathbf{U}$ is the vector of degrees of freedom (DoF), $\mathbf{M}$ is a modified global mass matrix with null entries corresponding to the pressure DoF, Δt is the time step size, $\mathbf{R}(\mathbf{U})$ is the residual vector, h is a numerical perturbation and $\mathbf{P}$ is the preconditioner. In this work a cheap and memory saving element-wise block-Jacobi preconditioner is employed. This approach becomes reliable as the time step decreases and the grid stretching at the wall is not excessive. The evaluation of this preconditioner costs nearly three times less than the incomplete LU factorization, ILU(0), of the full Jacobian matrix and, in terms of memory requirements, the resulting scheme is comparable to an explicit scheme. The CPU-time is further reduced via a lagging procedure for the preconditioner, which becomes more effective as the time step size decreases. Moreover, in order to avoid over solving the linear system, the relative tolerance of the solution is adaptively set one order lower than the solution accuracy in time, which is estimated using the embedded Runge–Kutta scheme.

3 Results

The computational domain employed is characterised by an aspect ratio $L/H = 8$, where L is the wall length and H is the distance between the two walls. Table 1 shows the computational details of various solutions obtained. The numbers of DoF employed are at least one order of magnitude lower than those of reference DNS [7], therefore the computations should be considered as ILES, apart from Case 1, at the lowest Rayleigh number, which is a DNS. Consistency relations, introduced in [7, 9], have been calculated on the entire domain and are reported in the table. It can be seen that as the resolution increases, enhancing the order of polynomial expansion or the grid size, the values become consistent with the numerical DNS reference Nu_R from [7]. An experimental correlation, $Nu_{exp} = 0.125Ra^{0.303}Pr^{0.25}$ from [8], has been also evaluated to confirm further the quality of the Implicit LES solution where the DNS data were not available. Clearly, it is more convenient, for what regards the number of DoF, to raise the polynomial order instead of the grid size (see, for instance, Cases 7 and 11 of Table 1). Note that for a well resolved simulation all the different definitions of the Nusselt number should be equivalent. However, since Nu_1, Nu_2 and Nu_4 involve the components of the gradients, which are one order less accurate, they are slower than Nu_3 to reach convergence.

Figure 1 shows the computed turbulent statistics using the time-space average operator $\langle\cdot\rangle_\pi$, which deals with quantities averaged on planes parallel to the walls. The DG-ILES solution fits well the DNS data despite having a number of DoF orders of magnitude lower than the reference, DoF_R. In particular, as the order k of the polynomial approximation $\mathbb{P}_k$ increases, the amount of numerical dissipation decreases, thus the peaks of fluctuation are described with a higher accuracy thanks also to the growth of the cut-off frequency of the scheme. This behaviour can be observed in Fig. 2 showing the power spectral density of the kinetic energy and temperature for the cases at $Ra = 10^6$. The spectra have been obtained using the time series of the variables in a probe point at $z = H/2$, inside the bulk region. Clearly, the resolved maximum frequency becomes broader by increasing the order of polynomial approximation (see, for instance, the blue and black solid lines of Fig. 2). However, due to the favourable spectral properties of the scheme, the resolved part of the inertial range remains more or less unaffected by the numerical resolution and fits very well the $-5/3$ (Kolmogorov scaling) and the $-7/5$ (Bolgiano scaling) laws for the kinetic energy and the temperature, respectively. These spectra are in agreement with those reported in [9]. A comparison between the solutions obtained at different time step sizes was performed raising eight times the Δt value (red and black lines of Fig. 2). It is clear that the resulting spectra are nearly superimposed both in the inertial range and in the dissipative range, where an additional numerical dissipation, although being very small, can be noticed. This result proves the computational benefits of using

Table 1 Computational details and computed Nusselt (Nu) numbers on various numerical setup of the DG-ILES solver. $\Delta t = \Delta\tilde{t}U_R/H$; $\tilde{t}$ is the dimensional time, $U_R = \sqrt{g\alpha\Delta\Theta H}$, g and α are the gravitational acceleration and the thermal expansion coefficient, $\Delta\Theta$ the temperature difference between the lower and upper walls. $\mathrm{DoF_R}$ is the number of DoF of the reference DNS [7]. $Nu_1 = 1 + \sqrt{RaPr}\langle\varepsilon\rangle_V$, $Nu_2 = \sqrt{RaPr}\langle\chi\rangle_V$, $Nu_3 = 1 + \sqrt{RaPr}\langle w\theta\rangle_V$, $Nu_4 = \langle Nu\rangle_w$; $\varepsilon = \sqrt{Pr/Ra}(\partial u_i/\partial x_j)(\partial u_i/\partial x_j)$ is the pseudo-dissipation of kinetic energy and $\chi = (1/\sqrt{RaPr})(\partial\theta/\partial x_i)(\partial\theta/\partial x_i)$ is the dissipation of the temperature squared. $\langle\cdot\rangle_V$ denotes the ensemble average of volume integrals, while $\langle\cdot\rangle_w$ indicates ensemble averages of wall-surface integrals. Nu_R is the reference Nusselt from [7]; $Nu_{exp} = 0.125Ra^{0.303}Pr^{0.25}$ from [8]

Case	Ra	$\mathbb{P}$	Grid	Δt	DoF	DoF_R/DoF	Nu_1	Nu_2	Nu_3	Nu_4	Nu_R	Nu_{exp}
1	10^5	5	$24 \times 24 \times 8$	0.1	258048	–	4.25	4.25	4.24	4.25	–	3.74
2		3	$16 \times 16 \times 8$	0.1	40,960	206.40	5.09	6.59	7.25	7.39		
3		4	$16 \times 16 \times 8$	0.1	71,680	117.94	6.07	7.18	7.71	7.56		
4		5	$16 \times 16 \times 8$	0.1	114,688	73.71	6.79	7.57	7.98	7.87		
5		6	$16 \times 16 \times 8$	0.1	172,032	49.14	7.30	7.84	8.15	8.10		
6	10^6	3	$20 \times 20 \times 10$	0.05	80,000	105.68	5.75	7.00	7.57	7.62	8.17	7.52
7		6	$20 \times 20 \times 10$	0.05	336,000	25.16	7.65	7.97	8.17	8.16		
8		6	$20 \times 20 \times 10$	0.1	336,000	25.16	7.65	7.97	8.17	8.16		
9		6	$20 \times 20 \times 10$	0.2	336,000	25.16	7.64	7.98	8.18	8.17		
10		6	$20 \times 20 \times 10$	0.4	336,000	25.16	7.56	7.96	8.16	8.15		
11		3	$32 \times 32 \times 16$	0.05	327,680	25.80	6.95	7.66	8.03	7.95		
12		3	$32 \times 32 \times 16$	0.05	327,680	228.70	8.39	12.18	13.77	14.05		
13		4	$32 \times 32 \times 16$	0.05	573,440	130.69	10.53	13.47	14.85	14.49		
14	10^7	5	$32 \times 32 \times 16$	0.05	917,504	81.68	12.09	14.22	15.34	15.14	15.55	15.11
15		6	$32 \times 32 \times 16$	0.05	1,376,256	54.45	13.17	14.66	15.46	15.50		
16		6	$40 \times 40 \times 20$	0.1	2,688,000	27.88	13.92	14.95	15.53	15.52		
17		3	$72 \times 72 \times 36$	0.1	3,732,480	–	15.46	23.94	27.57	27.64	–	
18	10^8	4	$72 \times 72 \times 36$	0.1	6,531,840	–	19.27	26.10	29.13	28.59	–	30.35
19		5	$72 \times 72 \times 36$	0.05	10,450,944	–	22.99	27.70	30.04	29.89	–	

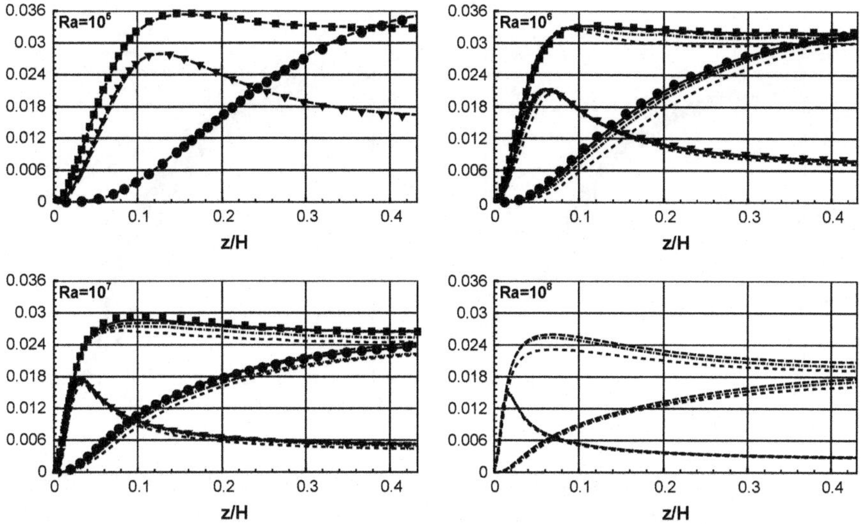

Fig. 1 Turbulent statistics of the Rayleigh–Bénard convection: $Ra = 10^5$ ($24 \times 24 \times 8$ mesh elements), $Ra = 10^6$ ($16 \times 16 \times 8$ mesh elements), $Ra = 10^7$ ($32 \times 32 \times 16$ mesh elements) and $Ra = 10^8$ ($72 \times 72 \times 36$ mesh elements). Reference DNS data symbols: kinetic energy fluctuations $\langle k' \rangle_\pi$ (*squares*), vertical velocity fluctuations $\langle w'^2 \rangle_\pi$ (*circles*) and temperature fluctuations $\langle \theta'^2 \rangle_\pi$ (*gradients*), from [7]. DG-ILES solutions at $\mathbb{P}_3$ (*dashed lines*), $\mathbb{P}_4$ (*dash dotted lines*), $\mathbb{P}_5$ (*long dashed lines*) and $\mathbb{P}_6$ (*solid lines*)

a high-order implicit time integration strategy. This is particularly true for an ILES computation where there exists a cut-off frequency due to the under-resolved space discretization. It is worth noting that in Fig. 2 the maximum frequency of the $\Delta t = 0.4$ spectrum is limited by the sampling frequency $1/\Delta t$ and could be increased using a dense-output algorithm for the implicit Runge–Kutta scheme. With this method, in fact, it is possible to evaluate an accurate solution within a time step. Note also that for $\Delta t > 0.4$ we observed a significant reduction of the accuracy and stability problems.

The developed solution strategy shows several advantages. First of all the high-order implicit time-stepping allows the use of very large time step sizes without introducing a significant dissipation to the one provided by the spatial discretization (see, in addition, the computed Nusselt number of Cases 7, 8, 9 and 10 of Table 1). Moreover, it reduces the memory footprint of the code (the saving is about 66% for a $\mathbb{P}_6$ approximation), and it allows a considerable CPU-time reduction if compared to a standard Matrix-Based algorithm, especially when the system is not ill conditioned (i.e. at small Δt). Regarding the last point, we can report that for the Case 7 of Table 1, the proposed approach saves about 84% of the CPU-time while for Case 10, characterized by the largest time step size here employed, the saving is still 33.3%. Finally, our implementation proves to be efficient and to scale optimally on highly parallel systems. For example, the full (starting from a trivial initial condition and

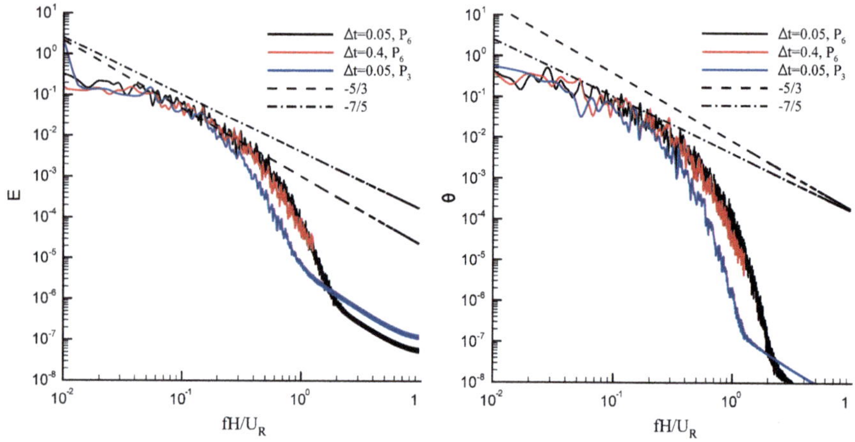

Fig. 2 Power spectral density of the Rayleigh–Bénard convection: $Ra = 10^6$ ($20 \times 20 \times 10$ mesh elements), kinetic energy E (*left*) and temperature θ (*right*)

up to the non dimensional time $t = 200$) $Ra = 10^8$ computation, involving more than 5.2×10^7 DoFs, is accomplished in less than one day using 68 Intel Xeon processors (1224 cores).

4 Conclusions

A matrix-free incompressible DG-ILES solver has been here applied to the solution of the Rayleigh–Bénard convection problems. Thanks to the favourable properties of the DG discretization, the code proves to be perfectly suited to the ILES of turbulent flows. The computed turbulent statistics and Nusselt numbers fits very well the reference DNS data and, for a fixed number of DoF, the quality of the solutions improves as the order of polynomial approximation raises. Consistently, the cut-off frequency of the scheme also increases, as it can be seen from the power spectral density of the solutions in the turbulent bulk region. Moreover, the high-order implicit time integration scheme allows the use of very large time steps without adding a notable numerical dissipation in the large scale motions, which are solved with an accuracy comparable with the one obtained at smaller time steps. The computational efficiency of the code has been further improved by the use of the matrix-free approach and of a lagged element-wise preconditioner. Future works will be devoted to assess the proposed solution strategy for the simulation of incompressible turbulent flows on complex geometries.

Acknowledgements We acknowledge the CINECA award, under the ISCRA initiative (Grant HP10B71OYM), for the availability of high performance computing resources and support.

References

1. G.J. Gassner, A.D. Beck, On the accuracy of high-order discretizations for underresolved turbulence simulations. Theor. Comput. Fluid Dynam. **27**(3–4), 221–237 (2013)
2. F. Bassi, L. Botti, A. Colombo, A. Crivellini, A. Ghidoni, F. Massa, On the development of an implicit high-order discontinuous galerkin method for dns and implicit les of turbulent flows. Eur. J. Mech. B/Fluids **55**, 367–379 (2016)
3. J. Rang, L. Angermann, New rosenbrock w-methods of order 3 for partial differential algebraic equations of index 1. BIT Numer. Math. **45**(4), 761–787 (2005)
4. A. Crivellini, F. Bassi, An implicit matrix-free discontinuous galerkin solver for viscous and turbulent aerodynamic simulations. Comput. Fluids **50**(1), 81–93 (2011)
5. F. Bassi, A. Crivellini, D. Di Pietro, S. Rebay, An implicit high-order discontinuous Galerkin method for steady and unsteady incompressible flows. Comput. Fluids **36**(10), 1529–1546 (2007)
6. F. Bassi, A. Crivellini, D.A. Di Pietro, S. Rebay, An artificial compressibility flux for the discontinuous Galerkin solution of the incompressible Navier–Stokes equations. J. Comput. Phys. **218**(2), 794–815 (2006)
7. R. Togni, A. Cimarelli, E. De Angelis, Physical and scale-by-scale analysis of Rayleigh–Bénard convection. J. Fluid Mech. **782**, 380–404 (2015)
8. H. Yang, Z. Zhu, Numerical simulation of turbulent Rayleigh–Benard convection. Int. Commun. Heat Mass Transfer **33**(2), 184–190 (2006)
9. R. Verzicco, R. Camussi, Numerical experiments on strongly turbulent thermal convection in a slender cylindrical cell. J. Fluid Mech **477**, 19–49 (2003)

Part V
Experiments

Measurement of Turbulent Spatial Structure and Kinetic Energy Spectrum—Part 1: Convection Record Method

Preben Buchhave and Clara M. Velte

Abstract A novel exact temporal to spatial mapping for point measurements in turbulence has been developed. The spatial record is obtained based on the instantaneous velocity magnitude, $u = |\mathbf{u}|$, creating an exact mapping between the sampling interval, Δt, and the spatial record counterpart, Δs, through the relation $\Delta s_n = u_n \Delta t_n$. n indicates the sample number in a measurement sequence. Summation of the consecutive streakline elements, Δs, corresponding to the convection distance of the fluid, results in a spatial "convection record". The exact mapping applies to all flows, since it is based on the instantaneous velocity magnitude, thereby incorporating all relevant aspects of the flow dynamics. Even high intensity non-equilibrium spatial records can be measured using this mapping, which is most straightforwardly applied using laser Doppler anemometry measurements. Computer simulated high intensity LDA data demonstrate the technique. The method will also be demonstrated on measurements in a round turbulent jet in part 2.

1 Introduction

The spatial structures of velocity fluctuations in turbulent flows are of great interest for theoretical understanding of turbulence and for realistic turbulence models for technological applications. Presently, there is particular interest in the statistical properties of unsteady and non-equilibrium turbulent flows where the classical Kolmogorov picture of a fully developed, equilibrium turbulence does not apply. Such flows may be either non-stationary or inhomogeneous and highly turbulent, which means that the traditional "Taylor's frozen turbulence hypothesis" cannot be used to

P. Buchhave (✉)
Intarsia Optics, Birkerød, Denmark
e-mail: buchhavepreben@gmail.com

C.M. Velte
Technical University of Denmark, Kgs. Lyngby, Denmark
e-mail: cmve@dtu.dk

R. Örlü et al. (eds.), *Progress in Turbulence VII*, Springer Proceedings in Physics 196, DOI 10.1007/978-3-319-57934-4_23

accurately convert the measured temporal record to a spatial one. In the following, we briefly describe the new time-to-space method (for a more detailed description, see [1]).

2 The New Time-to-Space Method

The traditional Taylor's Hypothesis is based on the assumption that the spatial fine-scale turbulent velocity structure is transported past the measurement point by the local mean velocity so quickly that the small scales do not have time to change, "so that an unchanging pattern of turbulent motion is swept past a stationary probe" (Taylor [2]). The temporal velocity record, t, can then be converted to a spatial one, s_{TH}, by means of the measured local mean velocity in the average flow direction, $\overline{u_1(\mathbf{x}_0, t)}$ say, by the formulas

$$\Delta s_{TH,n} = \overline{u_1(\mathbf{x}_0, t)} \Delta t_n \qquad \text{(sampling intervals)} \tag{1}$$

and

$$s_{TH,n} = \sum_{n'=1}^{n} \overline{u_1(\mathbf{x}_0, t)} \Delta t_{n'} \qquad \text{(spatial record)} \tag{2}$$

where $\mathbf{x}_0$ is the location of the measurement point, Δt_n is the n'th temporal sampling interval and Δs_n is the n'th spatial sampling interval. The velocity samples are the same measured ones, $u_i(s_n) = u_i(s_n)$, where $i = \{x, y, z\}$ indicates the measured velocity component, but the sampling intervals are different.

This method runs into problems when the turbulence intensity is greater than approximately 45%, a fact that was recognized early on [3–10] and studied in numerous publications over the years [4, 6, 8, 11–17]. The main problem is that the small spatial velocity structures are swept past the measurement point by the large, fluctuating convection velocity, which means that the temporal power spectrum cannot simply be converted to a spatial one by the scaling given by (2).

Our new method converts a temporal record into a spatial record, which we may name the "convection record", since it is based on the summation of small streak line elements, each corresponding to the convection distance of the fluid, Δs, during the temporal sampling interval, Δt, caused by the magnitude of the instantaneous velocity vector, $u_n = |\mathbf{u}_n| = \sqrt{u_x^2 + u_y^2 + u_z^2}$, see Fig. 1. The spatial seeding density should in practice approach the Nyquist sampling rate to avoid excessive noise in the spectrum. The insert shows the measurement volume, i.e. the volume in the flow from which Doppler signals are received. ΔA is the measurement cross section as seen from the flow direction and Δs is the spatial sampling interval.

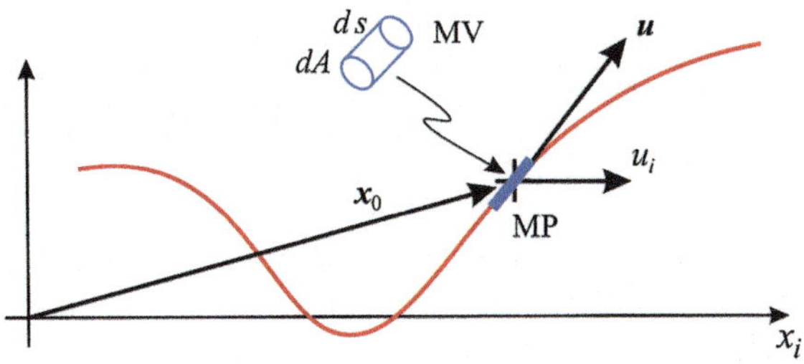

Fig. 1 The instantaneous streak line element through the measuring volume

The n'th convection distance element is given by

$$\Delta s_n = u_n \Delta t_n \qquad \text{(sampling intervals)} \tag{3}$$

and the spatial record is given by

$$s_n = \sum_{n'=1}^{n} u_{n'} \Delta t_{n'}. \qquad \text{(sampling record)} \tag{4}$$

The new spatial record now reflects the transport of spatial structures through the measuring volume, independent of the turbulence intensity. The velocity samples are the same, $u_i(s_n) = u_i(s_n)$, but the new spatial record now shows the true distance between spatial structures as they are convected past the measurement point.

In our work, we are particularly interested in the turbulence spectrum of the measured velocity component, u_x. The temporal and spatial energy spectra are given by (formulas provided without residence time weighting for clarity)

$$F_{u_x}(f) = \frac{1}{T}\tilde{u}_x(f)\tilde{u}_x^*(f) \quad \text{where} \quad T = \sum_{n=0}^{N-1} \Delta t_n \quad \text{is the temporal record length} \tag{5}$$

and

$$\tilde{u}_x(f) = \sum_{n=0}^{N-1} e^{-i2\pi f t_n} u_x(t_n) \Delta t_n \tag{6}$$

is the Fourier transform of the velocity component, $u_x(t)$. The spatial energy spectra are likewise given by

$$F_{u_x}(k) = \frac{1}{L}\tilde{u}_x(k)\tilde{u}_x^*(k) \quad \text{where} \quad L = \sum_{n=0}^{N-1} \Delta s_n \quad \text{is the spatial record length} \tag{7}$$

and

$$\tilde{u}_x(k) = \sum_{n=0}^{N-1} e^{-i2\pi k s_n} u_x(s_n)\Delta s_n \tag{8}$$

is the Fourier transform of the velocity component, $u_x(s)$.

Note, that the Fourier transform must be computed by discrete Fourier transform (DFT) since the sampling intervals are not equidistant. Moreover, in LDA measurements, the temporal sampling is a random (Poisson) process since the sampling times (arrival times, t_n) are determined by the arrival of seed particles to the measurement point. Thus, the transit times or residence times, δt_r, must be measured along with the arrival times and velocities to allow correction of the velocity—sampling rate correlation (also called velocity bias) by the residence time weighting method [18–21].

The method relies on the knowledge of the magnitude of the instantaneous velocity vector, $u_n = |\mathbf{u}_n| = \sqrt{u_x^2 + u_y^2 + u_z^2}$. However, with an LDA instrument that measures the residence time, the three velocity components need not be measured as the velocity magnitude can be estimated from knowledge of the measurement volume dimension in the flow direction, d_{MV} (measuring volume assumed spherical), and the residence time:

$$u_n = \frac{d_{MV}}{\delta t_n}. \tag{9}$$

Then the spatial record can be computed:

$$s_n = \sum_{n'=1}^{n} \frac{d_{MV}}{\delta t_{n'}} \Delta t_{n'} = \sum_{n'=1}^{n} \frac{d_{MV}}{\delta t_{n'}} (t_{n'} - t_{n'-1}) \tag{10}$$

where we have assumed the sampling rate so high that we can accept the time between samples as the sampling interval.

Since the spatial record, s_n, is one dimensional, and since the velocity vector is always in direction of s_n, the energy spectrum of u_n represents the total kinetic energy spectrum computed with a single, one-dimensional DFT.

$$F_u(k) = \frac{1}{L}\tilde{u}(k)\tilde{u}^*(k) \tag{11}$$

The final spectra are found as block averages of e.g. 100 records to reduce variance. The biggest source of variance in the final spectrum is the random sampling. The fluctuations in the particle path through the MV also cause variance in the spectral estimate, but this effect is smaller than the one due to random sampling.

3 LDA Simulation Results

To demonstrate the method, we performed laser Doppler anemometry (LDA) simulations based on a von Kármán spectrum (see yellow curve in RHS Fig. 2) [22]. This theoretical spectrum was fitted to a power spectrum from LDA measurements in a round turbulent jet, acquired at the centerline in the developed region (30 jet exit diameters downstream).

The spatial von Kármán spectrum was used as a filter to create a simulated turbulent velocity field that was sampled using a Poisson process to mimic the sampling process of the LDA. The spatial sampling bypasses the issue of velocity-sample rate bias and therefore a pure Poisson sampling without velocity modulation could be implemented. These fluctuations were convected through the simulated LDA measuring volume by a three-dimensional large scale Gaussian velocity field, to produce spatially and temporally varying convection velocities. The resulting turbulence intensity of the total signal was 54%. An average streamwise velocity of 1 ms^{-1} and a record length of 1 s was chosen to obtain a direct correspondence in the scaling of the spatial and temporal spectra.

The resulting randomly sampled time record produces an aliased time spectrum, Fig. 2 LHS. Residence time weighting has been implemented to avoid the velocity-sample rate bias inherent in LDA measurements [20]. The mapping back to spatial spectra is shown in Fig. 2 RHS using the convection record proposed herein as well as the classical Taylor's frozen field hypothesis. It is clear that the convection spectrum restores the von Kármán spectrum even in the presence of large temporal and spatial fluctuations, while Taylor's hypothesis is not able to restore in particular the small scales.

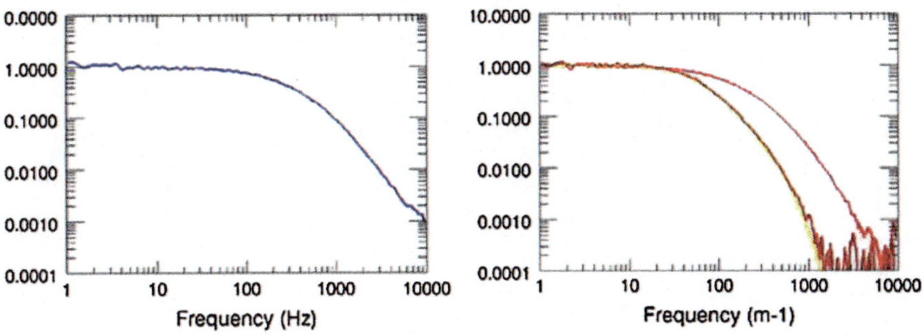

Fig. 2 Spectra of the von Kármán turbulence convected past the LDA measuring volume by a large low frequency Gaussian fluctuation. *Left*: *Blue* The temporal spectrum. *Right*: *Yellow* The original von Kármán model spectrum. *Dark red* The spatial spectrum restored. *Light red* The spatial spectrum restored by the conventional Taylor's hypothesis

4 Conclusions

The concept of convection records has been introduced and applied to computer simulations of LDA measurements from an LDA measurement in a turbulent round jet. The classical mapping of Taylor fails at the high turbulence intensity simulated, 54%, while the convection spectra are correctly mapped between space and time. This mapping may be implemented in any flow and can provide spatial spectra, correlations etc. even in inhomogeneous flows if the flow is stationary. Non-equilibrium flows of high turbulence intensities, where Taylor's hypothesis cannot be accurately invoked, is one of the central useful flows identified for implementing this method.

References

1. P. Buchhave, C.M. Velte, Conversion of measured turbulence spectra from temporal to spatial domain. in *Whither Turbulence and Big Data in the 21st Century?* (Springer International Publishing, 2017), pp. 343–362
2. G.I. Taylor, The spectrum of turbulence. Proc. R. Soc. Lond., Ser. A **164**, 476 (1938)
3. C.C. Lin, On Taylor's hypothesis and the acceleration terms in the Navier-Stokes equations. J. App. Math. **10**, 295 (1953)
4. J.L. Lumley, Interpretation of time spectra measured in high-intensity shear flows. Phys. Fluids **8**, 1056 (1965)
5. M.J. Fisher, P.O.A.L. Davies, Correlation measurements in a nonfrozen pattern of turbulence. J. Fluid Mech. **18**, 97–116 (1964)
6. J.C. Wyngaard, S.F. Clifford, Taylor's hypothesis and high frequency turbulence spectra. J. Atm. Sci. **34**, 922 (1977)
7. A.S. Gurvich, Influence of the temporal evolution of turbulent inhomogeneities on frequency spectra. Atmos. Ocean. Phys. **16**, 231–237 (1980)
8. R.A. Antonia, N. Phan-Thien, A.J. Chambers, Taylor's hypothesis and the probability density functions of temporal velocity and temperature derivatives in a turbulent flow. J. Fluid Mech. **100**, 193 (1980)
9. J.W. Deardorff, G.E. Willis, Investigation of the frozen-turbulence hypothesis for temperature spectra in a convectively mixed layer. Phys. Fluids **25**, 21–28 (1982)
10. J.C. Kaimal, R.A. Eversole, D.H. Lenschow, B.B. Stankov, P.H. Kahn, J.A. Businger, Spectral characteristics of the convective boundary layer over uneven terrain. J. Atmos. Sci. **39**, 1098–1114 (1982)
11. G. Heskestad, A generalized Taylor hypothesis with application for high Reynolds number turbulent shear flows. J. Appl. Math. **32**, 735 (1965)
12. H. Tennekes, Eulerian and Lagrangian time microscales in isotropic turbulence. J. Fluid Mech. **67**, 561 (1975)
13. W.C. Thacker, A transformation relating temporal and spatial spectra of turbulent kinetic energy. Boulder, Colo. : Dept. of Commerce, National Oceanic and Atmospheric Administration, Environmental Research Laboratories, Atlantic Oceanographic and Meteorological Laboratories, Miami, Florida (1977)
14. F.H. Champagne, The fine-scale structure of the turbulent velocity field. J. Fluid Mech. **78**, 67 (1978)
15. K.B.M.Q. Zaman, A.K.M.F. Hussain, Taylor's hypothesis and large-scale coherent structures. J. Fluid Mech. **112**, 379 (1981)
16. J. Mi, R.A. Antonia, Corrections to Taylor's hypothesis in a turbulent circular jet. Phys. Fluids **6**, 1548 (1994)

17. E. Gledzer, On the Taylor hypothesis corrections for measured energy spectra of turbulence. Physica D **104**, 163 (1997)
18. P. Buchhave, Errors and correction methods in turbulence measurements with the LDA. PhD Dissertation, State University of New York at Buffalo (1979)
19. C.M. Velte, Characterization of vortex generator induced flow. PhD Dissertation, Technical University of Denmark (2009)
20. C.M. Velte, W.K. George, P. Buchhave, Estimation of burst-mode LDA power spectra. Exp. Fluids **55**, 1674 (2014)
21. P. Buchhave, W.K. George, J.L. Lumley, The measurement of turbulence with the laser-Doppler anemometer. Ann. Rev. Fluid Mech. **11**, 443 (1979)
22. P. Buchhave, C.M. Velte, Reduction of noise and bias in randomly sampled power spectra. Exp Fluids **56**, 79 (2015)

Measurement of Turbulent Spatial Structure and Kinetic Energy Spectrum—Part 2: Convection Record Measurements

Clara M. Velte, Preben Buchhave and Azur Hodžić

Abstract A novel exact temporal to spatial mapping for point measurements in turbulence has been applied to various flow conditions existing in a round turbulent jet. The conditions range between equilibrium and non-equilibrium as well as mid to high turbulence intensities. The exact mapping applies to all flows, including high intensity non-equilibrium flows, since it is based on the instantaneous velocity magnitude, thereby incorporating all relevant aspects of the flow dynamics. Development of the jet turbulence along the stream, from non-equilibrium to equilibrium, is observed. In the developed region of the jet, Taylor's hypothesis is tested and the spectra using the novel exact mapping is validated with excellent agreement against directly measured spatial spectra in a mapped similarity space using PIV. The method is observed to produce the expected results even at turbulence intensities of the order of 450%.

1 Introduction

In the recent developments in turbulence research, focus has lately been directed towards non-equilibrium flows. These flows do not follow the classical Kolmogorov picture of a universal equilibrium range in which the smallest scales can be regarded to adapt instantly to surrounding conditions and therefore be considered in a state of (quasi-) equilibrium. Taylor's frozen field hypothesis has typically been invoked for measuring spatial records from temporal records. Unfortunately, this low-order mapping is known to fail in particular for the most interesting high-intensity flows.

C.M. Velte (✉) · A. Hodžić
Technical University of Denmark, Kgs. Lyngby, Denmark
e-mail: cmve@dtu.dk

A. Hodžić
e-mail: azuhod@mek.dtu.dk

P. Buchhave
Intarsia Optics, Birkerød, Denmark
e-mail: buchhavepreben@gmail.com

R. Örlü et al. (eds.), *Progress in Turbulence VII*, Springer Proceedings
in Physics 196, DOI 10.1007/978-3-319-57934-4_24

In this article, we describe the application of a new method (see part 1 or [1] for more details) for converting temporal records acquired at a fixed measurement point with a laser Doppler anemometer (LDA) to a spatial record reflecting the transport or convection of spatial structures past the measuring point. This mapping is completely general and can therefore be applied to any flow, whether time dependent, inhomogeneous or even transitional.

Since the velocity data are acquired at a single point, the flow field does not have to be homogeneous to allow computation of statistical quantities such as correlations, spectra and structure functions. However, in order that spectra may be computed from the temporal record, the flow must be stationary within the measured temporal record length.

2 Measurements

Measurements were performed in two separate experiments with two identical round jets in air with exit diameter $D = 10\,\text{mm}$, one using an LDA [2] and the other using a particle image velocimeter (PIV) [3, 4], see [5, 6] for a detailed description of the flow facility. The turbulent jet is an ideal test bed for exploration of turbulence in both non-equilibrium and equilibrium regions since it is well known theoretically as well as from experiments (see e.g. Hinze 1992 [7]). Moreover, the logarithmic mapping of velocity components along the jet axis allows PIV measurements to be performed in two local homogeneous directions [8]. We used an in-house LDA (Fig. 1), designed to give high spatial resolution with a near spherical measurement volume (with diameter $d_{MV} = 200\,\mu\text{m}$) and high S/N using separate emission optics and receiving optics at 90° to each other. We used digital sampling with a high sample rate (4 channels up to 2 GS/s per channel) and high resolution (12–14 bits). Processing was performed with highly transparent and flexible software to allow optimum adjustment of filter

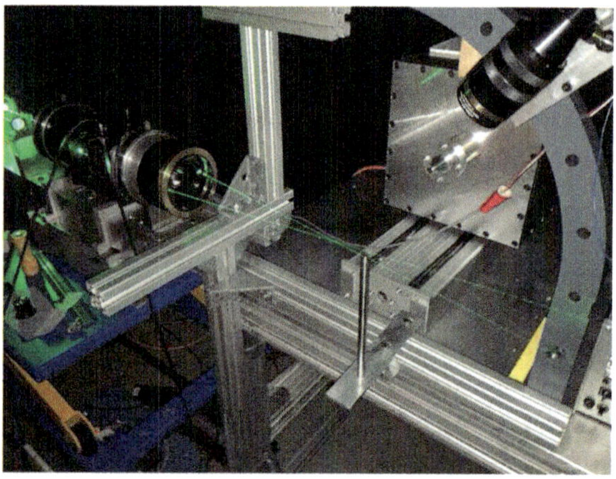

Fig. 1 Jet experiment showing jet orifice, LDA sender and LDA receiver positioned at an angle. Measurements in the current study were made with the detector positioned at 90° to the laser beams

settings and velocity validation. The whole system is optimized for highest possible S/N to allow measurements of spectra with high dynamic range. The far-field PIV measurements were obtained from the data of [3, 4] where the experimental setup is described in detail. The data was sampled using a two-component PIV in the longitudinal direction of the jet with two Dantec FlowSense EO 16 $MPix$ cameras and a pixel pitch of 7.4 μm. 60 mm Nikon lenses were used with an aperture of $f = 2.8$. The two cameras were combined to extend the total view of the far-field reducing the effect of windowing on the spectra. The LDA and PIV measurements were performed at $Re = 20{,}000$. The LDA spectra were based on 100 records with a 1 s record length and an average sampling frequency of approximately 5000 S/s. The PIV measurements from [3, 4] were based on 11,000 independent realizations in the developed region interval $x = 30 - 100D$.

3 Developed Jet Region

Figure 2a shows spatial spectra measured in the developed region $x = 30 - 100D$. These spectra are obtained from PIV measurements where the similarity scaling

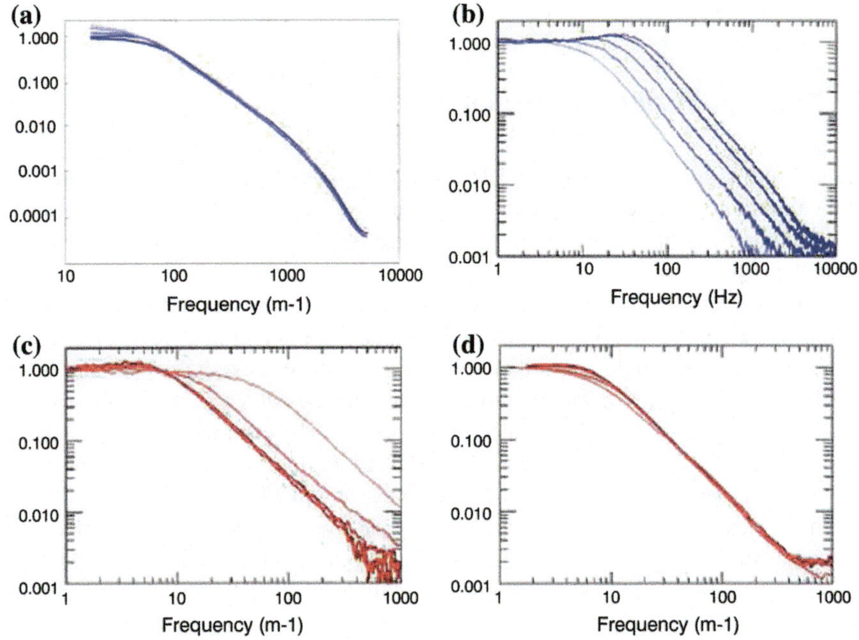

Fig. 2 Measured spectra from the fully developed region of the turbulent jet at $x = 30D$ at different radial positions (*heavy to light color* increasing distance from axis). All spectra are normalized with their respective mean square values for the sake of comparison. **a** Spatial spectra from PIV measurements. **b** Temporal spectra measured with LDA. **c** Spatial spectra obtained by invoking Taylor's hypothesis. **d** Spatial spectra from the same LDA data, converted with the new method

from [8] is applied in order to map the field to similarity coordinates. The coordinate transformation imposes homogeneity in the streamwise direction of the flow allowing one-dimensional spatial spectra to be extracted at each radial position in similarity space. The PIV spectra are normalized by their respective mean square energies, indicating that the spatial structure is uniform across the jet in this interval of the developed region.

Figure 2b shows LDA temporal spectra, acquired at $x = 30D$, at radial coordinates $r = 0,\ 13,\ 26,\ 39,\ 52$ mm, normalized to $u'^2_x = 1$. These spectra show the effect of the convection as the energy is shifted to higher frequencies closer to the axis where the average convection velocity is highest. Invoking Taylor's hypothesis to obtain the spatial energy spectra and again normalizing the spectra by their mean square values, Fig. 2c, reveal that the computed spectra do not collapse to the same form in regions of high turbulence intensity, deeming Taylor's hypothesis inappropriate particularly in the outer parts of the jet. In the outermost radial position, corresponding to two jet half-widths, the turbulence intensity is about 450%. The spatial spectra in Fig. 3d, computed by the convection record result in a near perfect collapse of the curves using the same LDA data, in agreement with the spatial spectra directly measured using PIV in Fig. 2a. This demonstrates the validity of the new conversion method, even at 2.5 times the jet-half width, where the signal displays high intermittency and turbulence intensity.

4 Non-equilibrium Jet Region

Figure 3 shows measurements with LDA in two regions of the jet, where equilibrium of the turbulent cascade has not developed. The spectra in Fig. 3a, b are measured at $x = 15D$ and Fig. 3c, d show spectra measured at $x = 10D$. The temporal spectra show the expected displacement due to the increasing velocity as we approach the axis. However, the spatial spectra do not collapse as we saw in the fully developed region. We also note that the shapes of the spectra are not similar, revealing subtle differences in the spatial structure of the velocity at different radial positions. Noteworthy is perhaps also the fact that the general tendency at $x = 15D$ is still a slope of $-5/3$ as expected only in the fully developed jet. The even less developed spectra displayed in Fig. 3c, d show the same features, perhaps even more pronounced. However, these spectra do not show a frequency range with a constant slope.

These measurements are only preliminary, and we plan to make more detailed studies of the non-equilibrium regions of the turbulent jet involving also structure functions and dissipation measurements using the new spatial conversion method.

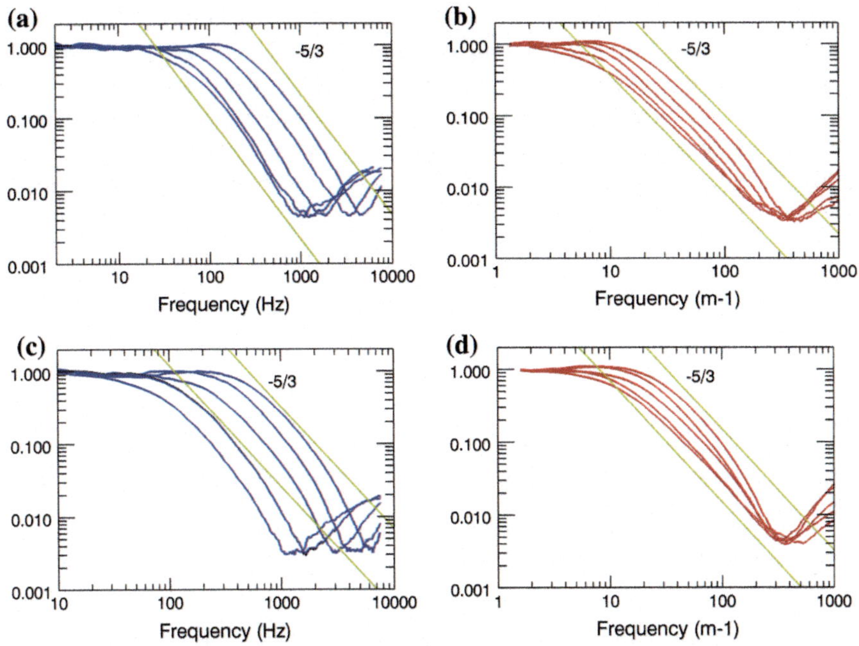

Fig. 3 Temporal spectra (*blue*) and spatial spectra (*red*) from non-equilibrium regions of the turbulent jet measured with LDA at various radial positions. **a**, **b** Temporal and spatial spectra using the new conversion method measured at $x = 15D$. **c** and **d** same measured at $x = 10D$

5 Conclusions

A new temporal-to-spatial record conversion method not hampered by the approximations inherent in Taylor's Hypothesis has been applied to LDA measurements in a round, turbulent jet in air. The LDA measurements are compared to direct spatial spectra measurements in an identical jet using PIV and a homogenizing similarity mapping. Both temporal spectra based on the randomly sampled temporal records and spatial spectra from spatial records converted both by the new method and by the traditional Taylor's Hypothesis were computed. The PIV measurements show that the spatial spectra measured in the fully developed region of the jet when normalized to the same mean square value are identical. The spectra computed from spatial records converted from the temporal LDA records show near perfect collapse in agreement with the PIV spectra when the new method is used. Taylor's Hypothesis provides collapsing spectra only in regions with turbulence intensity less than 45% while failing totally in the outer region of the jet.

Measurements were also conducted with LDA in regions nearer to the exit, where the turbulence is not fully developed. Interesting features due to the non-equilibrium were revealed. These measurements cannot be made with PIV because the flow cannot be considered locally homogeneous and thus statistical results do not have any physical meaning. Hot-wire measurements cannot be made accurately because of the high turbulence intensity. LDA measures temporal records that can be converted by Taylor's Hypothesis or by our new method. Our results show that only our new method provides identical spatial spectra as expected from PIV data when applied to the fully developed region of the jet. This leaves LDA the only option for measurements in the high turbulence intensity regions of the non-equilibrium parts of the jet, where local homogeneity cannot be expected.

References

1. P. Buchhave, C.M. Velte, Conversion of measured turbulence spectra from temporal to spatial domain. in *Whither Turbulence and Big Data in the 21st Century?* (Springer International Publishing, Berlin, 2017), pp. 343–362
2. C.M. Velte, W.K. George, P. Buchhave, Estimation of burst-mode LDA power spectra. Exp. Fluids **55**, 1674 (2014)
3. A. Hodžić, PIV measurements on a turbulent free jet-spatial decomposition of a turbulent free jet using proper orthogonal decomposition. MSc Dissertation, Technical University of Denmark (2014)
4. A. Hodžić, C.M. Velte, K.E. Meyer, POD mode robustness for the turbulent jet sampled with PIV. In *Whither Turbulence and Big Data in the 21st Century?* (Springer International Publishing, Berlin, 2017), pp. 163–188
5. D. Jung, S. Gamard, W.K. George, Downstream evolution of the most energetic modes in a turbulent axisymmetric jet at high Reynolds number. Part 1. The near-field region. J. Fluid Mech. **514**, 173–204 (2004)
6. S. Gamard, D. Jung, W.K. George, Downstream evolution of the most energetic modes in a turbulent axisymmetric jet at high Reynolds number. Part 2. The far-field region. J. Fluid Mech. **514**, 205–230 (2004)
7. J.O. Hinze, *Turbulence: An Introduction to Its Mechanism and Theory* (McGraw Hill, New York, 1959)
8. D. Ewing, B. Frohnapfel, W.K. George, J.M. Pedersen, J. Westerweel, Two-point similarity in the round jet. J. Fluid Mech. **577**, 309 (2007)

Comparison of the Development of a Wind Turbine Wake Under Different Inflow Conditions

Ingrid Neunaber, Jannik Schottler, Joachim Peinke and Michael Hölling

Abstract We present a wind tunnel study of a model wind turbine wake with regard to different inflow conditions. The aim is to examine the influence of intermittent flows on the wake. For this, a regular grid that produces non-intermittent inflow and an active grid that creates intermittent inflows have been used. As a reference case, also laminar inflow conditions were examined. The development of the centerline turbulence intensity and mean velocity with increasing distance from the wind turbine was investigated, and an influence of the different characteristic features of turbulence on the wake recovery was found. The results expand and partially contradict previous results.

1 Introduction

Turbulence is shown to have a major influence on the wind energy conversion process. As wind energy converters are usually built in wind farms, not only the atmospheric turbulence influences the machines, but also the turbulence that develops behind the wind energy converters. To optimize the wind farm layout and to design wind energy converters according to the surrounding conditions, profound knowledge of the wind turbine wake and its development is important. A first step towards a proper description is to scrutinize the development of the velocity mean $\bar{v}$ and the standard deviation σ_v, which has already been done in several CFD, laboratory and field studies. Here, a study based on wind tunnel experiments is presented, where the focus is on the examination of the influence of different turbulent inflows on the development of the wake of a load-controlled model wind turbine.

In the past, several laboratory studies of wakes of model wind turbines exposed to turbulent inflow conditions have been carried out. In [1], the wake of a model wind turbine in two different turbulent boundary layers is examined, namely, the influence of a rough and a smooth boundary layer on different quantities in the wind

I. Neunaber (✉) · J. Schottler · J. Peinke · M. Hölling
ForWind, Institute of Physics, University of Oldenburg,
Küpkersweg 70, 26129 Oldenburg, Germany
e-mail: ingrid.neunaber@uni-oldenburg.de

R. Örlü et al. (eds.), *Progress in Turbulence VII*, Springer Proceedings
in Physics 196, DOI 10.1007/978-3-319-57934-4_25

turbine wake at various downwind positions are compared, and wake profiles are studied. They conclude that the inflow turbulence is beneficial for the recovery of the velocity deficit in the wake (in the following called wake recovery). In comparable studies like e.g. [2], the effect that turbulent inflow leads to a faster wake recovery is confirmed.

Above-mentioned studies come to the general conclusion that the inflow turbulence has an influence on the wake. However, in many studies, including [1, 2], the inflow turbulence is only specified by mean value and standard deviation. Consequently, the intermittency, which is one important feature of the atmospheric wind (cf. [8]) and which could be related to an increased wind turbine failure rate [7], is often not accounted for in the layout and discussion of experiments. The term intermittency refers to the gustiness of a flow, and it can be characterized by the probability density function (PDF) $p(\delta v(\tau))$ of velocity increments $\delta v(\tau) = v(t+\tau) - v(t)$. This description takes into account the velocity fluctuations over a time lag τ. Here, it should be pointed out that the time lag τ or, respectively, the corresponding spacial scale obtained by the use of Taylor's hypothesis of frozen turbulence, plays an important role for the wake structure if these scales are within those that determine the wake structure, i.e. the size of the rotor or the size of the chord of the blades.

The aim of this paper is to examine the influence of intermittency on the wake. For this, the impact of two fundamentally different turbulent inflows, one with intermittent and one with non-intermittent features on the relevant scales, is examined experimentally. In Sect. 2, the experimental setup is presented. Section 3 shows the results which are directly discussed. Finally, in Sect. 4, a conclusion closes this article.

2 Experimental Setup

The setup is presented in Fig. 1. A model wind turbine with rotor diameter $D = 58$ cm and tip speed ratio $\lambda = 7.2$ is installed in a closed-loop wind tunnel with open test-section [6]. The measurements of the wake are carried out with seven hot-wire probes aligned in an array that is traversed along the centerline up to a distance of $X/D = 4.65$. The sampling frequency is $f_s = 20$ kHz, and a low-pass filter with a cut-off frequency of 10 kHz is used.

To create the above-mentioned different turbulent inflows, a regular and an active grid are used. The regular grid (mesh size of 4.4 cm) creates on the scales of the rotor non-intermittent, i.e. Gaussian, turbulence, while the active grid is capable of creating respective intermittent flows (cf. e.g. [8]): Several vertical and horizontal axes with diamond-shaped flaps can be rotated, and a motion pattern of all axes allows to create customized turbulent flows [3]. In this experiment, we use a motion protocol that recreates typical statistical characteristics of free field wind data, rescaled to wind tunnel dimensions. As a third reference case, a quite laminar inflow is used.

For the two turbulent inflow conditions, the PDFs of velocity increments are plotted in Fig. 2 for different time lags τ. The PDFs are normalized to the standard deviations of the increment time series, σ_τ, and they are shifted vertically for clarity.

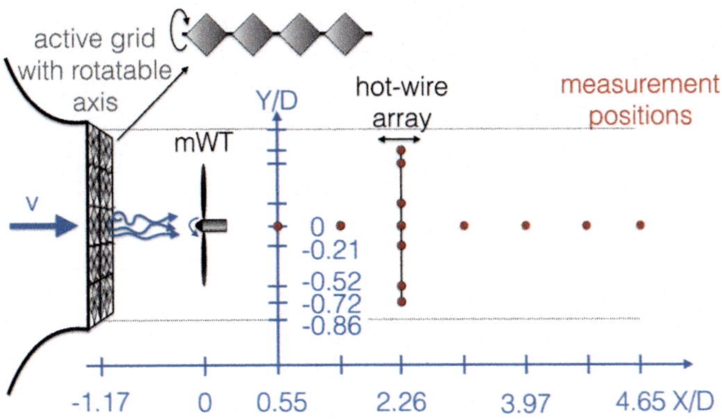

Fig. 1 Top-view of the setup: The model wind turbine is placed in different turbulent flows created by a regular and an active grid. The wind tunnel with outlet (80×100) cm^2 is operated with an open test section. A hot-wire array with seven probes is used to traverse the centerline of the wake

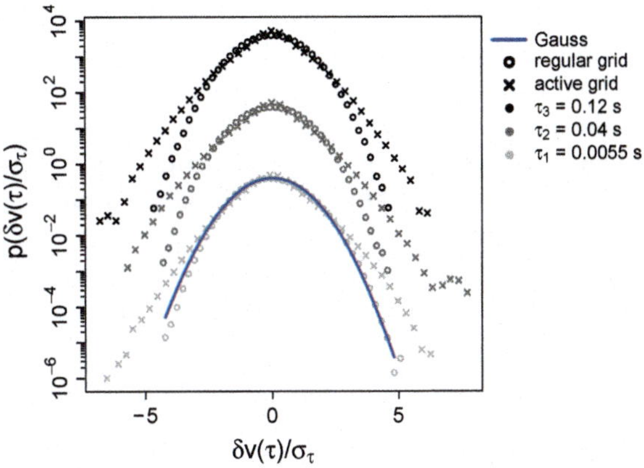

Fig. 2 PDF of velocity increments for different time lags τ. $\tau = 0.0055$ s corresponds to the chord length at approximately 50% of the radius, $\tau = 0.04$ s corresponds to the rotor radius and $\tau = 0.12$ s corresponds to 1.5 D. The PDFs are shifted vertically for clarity

The standard deviations of the active grid $\sigma_{\tau,act}$ are for the chosen τ twice the value of the regular grid's standard deviations $\sigma_{\tau,reg}$. The corresponding spatial scales are chosen to match the turbine geometries: $1.5D$ corresponds to $\tau = 0.12$ s ≈ 0.87 m, $0.5D$ corresponds to $\tau = 0.04$ s ≈ 0.28 m and the chord length at approximately 50% of the radius corresponds to $\tau = 0.0055$ s ≈ 0.04 m. It can clearly be seen that the regular grid produces Gaussian statistics, whereas the active grid creates intermittent statistics on our selected scales.

The mean inflow velocities and turbulence intensities $TI = \sigma/\bar{v}$ of the free inflows (i.e. without turbine) at rotor position are presented in Table 1. The values are averaged over all seven sensors. In the following, the velocities are used as reference and normalization inflow velocities. As the mean wind speeds are set by a constant wind tunnel motor control voltage, the inflow velocity varies for the different inflow conditions.

Table 1 Mean averaged velocities $\bar{v}_0$ and turbulence intensities $TI_0 = \sigma_0/\bar{v}_0$ of the different inflow conditions at rotor position (no turbine, thus the index 0)

	Laminar	Regular grid	Active grid
$\bar{v}_0$/m/s	7.56	7.28	8.07
TI_0/%	1.36	6.72	12.81

3 Results and Discussion

In the following, the downstream evolution of the normalized mean velocity $\bar{v}/\bar{v}_0$ and turbulence intensity TI/TI_{peak} in the wake are shown and discussed with regard to the inflow conditions (cf. Fig. 3a, b).

The mean velocity drops as expected due to the pressure gradient caused by the turbine. At $X/D \approx 2$, the wake recovery starts. An influence of the inflow turbulence is visible: In case of the laminar and the intermittent inflow, the maximum wake deficit and the recovery are comparable. The maximum wake deficit in case of the non-intermittent inflow is significantly higher, but the wake recovery appears to be faster, although $TI_{0,reg}$ is roughly half of $TI_{0,act}$. Therefore, contradictory to other studies, we find that a higher turbulence degree in the inflow conditions does not necessarily lead to a faster recovery of the mean wind speed. The statistic characteristics of the inflow turbulence on turbine-related scales seem to play an additional role in the wake development.

In Fig. 3b, the development of the turbulence intensity over X/D is shown. Inspired by [4], the turbulence intensity is normalized to the respective peak turbulence

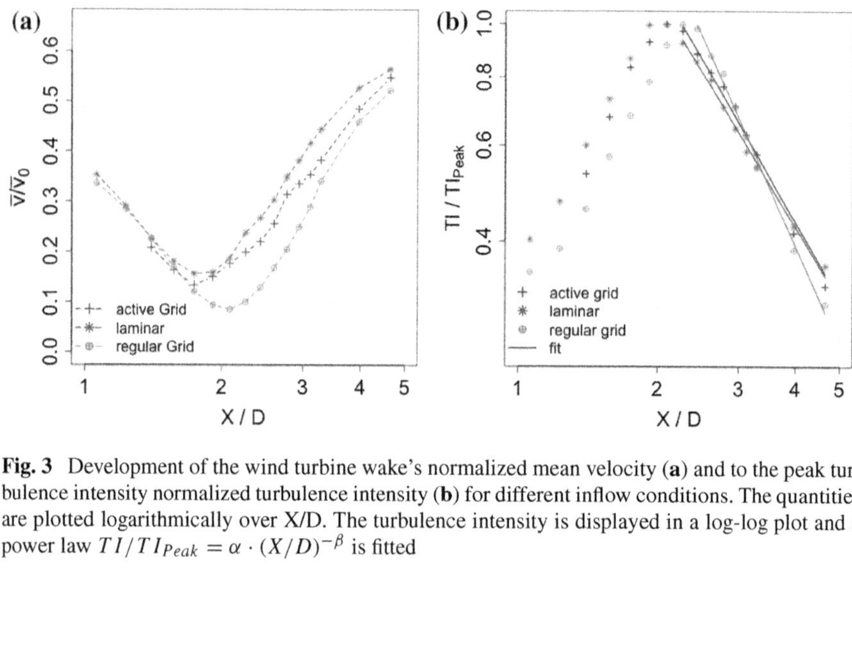

Fig. 3 Development of the wind turbine wake's normalized mean velocity (**a**) and to the peak turbulence intensity normalized turbulence intensity (**b**) for different inflow conditions. The quantities are plotted logarithmically over X/D. The turbulence intensity is displayed in a log-log plot and a power law $TI/TI_{Peak} = \alpha \cdot (X/D)^{-\beta}$ is fitted

Table 2 Fit parameters of the TI's power law fit $TI/TI_{Peak} = \alpha \cdot (X/D)^{-\beta}$

	Laminar	Regular grid	Active grid
$\alpha \pm \Delta\alpha$	2.87 ± 0.12	5.21 ± 0.38	3.23 ± 0.20
$\beta \pm \Delta\beta$	1.37 ± 0.04	1.87 ± 0.07	1.45 ± 0.07
χ^2_{PL}	0.00015	0.00028	0.00037

intensity TI_{peak} in an attempt to collapse the curves. It can be seen that the turbulence intensity peaks around X/D$\approx$ 2 and then decays. As publications as [5] have shown a power law decay of the velocity deficit, which is directly linked to the turbulence intensity, a power law $TI/TI_{Peak} = \alpha \cdot (X/D)^{-\beta}$ was fitted to the decay region. The fit parameters and the mean square residuals χ^2 can be found in Table 2. The low values of χ^2 indicate that the power law fit shows a good agreement.

To our knowledge, we are the first ones to connect the turbulence decay to classical wake decay description methods and to investigate also the evolution of the turbulence intensity. In analogy to the behavior of the mean velocity, β is similar in case of the laminar and the intermittent inflow, but differs in case of non-intermittent inflow. As a consequence of the higher decay exponent, in case of the non-intermittent inflow, the turbulence intensity decays faster, and thus the decay curve even crosses the other curves. This indicates that the intermittent turbulence counteracts the beneficial faster turbulence decay in case of Gaussian inflow turbulence.

4 Summary and Conclusion

We presented a study of the influence of different inflow conditions on the development of a wind turbine wake. Hot-wire measurements have been carried out at different positions along the streamwise axis up to $X/D = 4.65$.

The evolution of the mean velocity at centerline is shown to be dependent on the statistical characteristics of the inflow turbulence. While the wake recovers similarly for laminar and intermittent inflow conditions, the wake recovery in case of non-intermittent turbulence is faster, although the inflow turbulence was significantly lower compared to the intermittent case.

The decay of the turbulence intensity can be approximated by a power law fit after peaking around $X/D \approx 2$. Similar to the mean velocity, the inflow conditions also affect the turbulence decay. The turbulence intensity decays faster in case of non-intermittent inflow although having the lower inflow turbulence degree.

Both analyses suggest that on turbine scales intermittent turbulence counteracts the recovery of the mean velocity and the decay of the turbulence intensity. Whether this trend is conserved at larger distances X/D has to be further examined.

In conclusion, an influence of different characteristic features of the inflow turbulence on the wake of a model wind turbine could be shown. Therefore, a description of the inflow with mean velocity and turbulence intensity is not sufficient, and experiments in turbulent inflows have to be designed carefully to gain profound knowledge in cases of turbulence generation in turbulent surroundings.

Acknowledgements This work is funded by the Federal Environmental Foundation (DBU), Germany.

References

1. L.P. Chamorro, F. Port-Agel, A wind-tunnel investigation of wind-turbine wakes: boundary-layer turbulence effects. Bound. Layer Meteor. **132**, 129–149 (2009)
2. Y. Jin et al., Effects of freestream turbulence in a model wind turbine wake. Energies **9**, 830 (2016)
3. P. Knebel, A. Kittel, J. Peinke, Atmospheric wind field conditions generated by active grids. Exp. Fluids **51**, 471–481 (2011)
4. N. Mazellier, J.C. Vassilicos, Turbulence without Richardson–Kolmogorov cascade. Phys. Fluids **22**, 075101 (2010)
5. V.L. Okulov et al., Wake effect on a uniform flow behind wind-turbine model. J. Phys.: Conf. Ser. 625, 012011 (2015)
6. J. Schottler et al., Design and implementation of a controllable model wind turbine for experimental studies, J. Phys.: Conf. Ser. **753**, 072030 (2016)
7. P. Tavner et al., The correlation between wind turbine turbulence and pitch failure, in *Proceedings of EWEA 2011* (2011)
8. M. Wächter et al., The turbulent nature of the atmospheric boundary layer and its impact on the wind energy conversion process. J. Turbul. **13**, N26 (2012)

The Development of Packets of Hairpin Vortices in Laminar Channel Flows in Response to Localized Disturbances

Jun-De Li

Abstract We report the responses of laminar channel flows to localized disturbances at different Reynolds numbers and initial perturbation amplitudes by using direct numerical simulations (DNS). It is found that, when the Reynolds number and the local perturbation are above some critical values, the local disturbance generates a strong and long low speed streak among many weaker ones together with a pair of streamwise vortices next to the strong low speed streak. The spatiotemporal development of the flow structures shows that the first hairpin vortex appears above and at the middle of the strong low speed streak and several secondary hairpin vortices then appear behind and ahead of the primary hairpin vortex, and thus forming a packet of hairpin vortices. It is found that the hairpin vortices seem to be generated by the Kevin–Helmhotz type instability, and some of the hairpin vortices are formed from pairing two adjacent smaller hairpin vortices generated from the roll up of shear layer near the streamwise vortex pair. It is also found that the variance of the vertical velocity is a better indicator for showing the start of nonlinear effect.

1 Introduction

Bypass transition from laminar flow to instabilities and turbulence under the localized three-dimensional disturbance has long been considered [1] as an alternative to the two-dimensional Tollmien–Schlichting (TS) waves, and has been studied extensively both numerically and experimentally [2–4]. It is generally believed that for a laminar shear flow constrained by at least one solid wall, when the Reynolds number and the amplitude of the localized disturbance are above their respective critical values, a transient energy growth due to the "lift-up" effect [5] will generate sufficient amplitude in fluctuating velocity to trigger the nonlinear development of the disturbances

J.-D. Li (✉)
College of Engineering and Science, Victoria University, Melbourne, Australia
e-mail: Jun-De.Li@vu.edu.au

J.-D. Li
Institute for Turbulence-Noise-Vibration Interactions and Control,
Shenzhen Graduate School, Harbin Institute of Technology, Shenzhen, China

R. Örlü et al. (eds.), *Progress in Turbulence VII*, Springer Proceedings
in Physics 196, DOI 10.1007/978-3-319-57934-4_26

and result in the transition to turbulence. The growth of the disturbance is generally measured using the fluctuation energy which includes contributions from all the three velocity components. The indication of nonlinear triggering is normally provided by the fact that this energy grows faster and would not decay in a full simulation (or measurements) than that from a linearised analysis (simulation). In this paper, we will show that this fluctuation energy (or the streamwise velocity variance) may not be a sensitive indicator for identifying the start of such nonlinear effect.

In [4], the development of the laminar channel flow fields under localized disturbances was investigated and it was found that with large enough initial perturbations, the flow development reaches the stage of a small turbulent spot after a relatively short time. The appearance of the spot has been considered as a precursor for the fully developed turbulence. It was shown in [4] that due to strong nonlinearity, the flow develops some streamwise vortex whose breakdown results in a rollup process. Although it was conjectured that this rollup process produces some lambda vortex structures similar to that from the secondary instability process [6], the detailed rollup process and its consequence are not clear. Here the detailed vortex structures from this process are looked in details. It seems that a vortex packet consisting of several hairpin vortices formed sequentially can be developed and the overall structures is similar to the packet of hairpin vortices in fully turbulent channel flows as found in [7]. This shows that the formation of packet of hairpin vortices may be a common feature for laminar to turbulence transition and fully developed turbulence.

2 Direct Numerical Simulation

The DNS simulations were based on the well-known KMM scheme [8] and the spectrally accurate Galerkin method with basis functions consisting of several Chebyshev polynomials and satisfying the homogeneous boundary conditions for the normal to the wall vorticity and velocity, respectively. This method results in better conditional numbers numerically in comparison with the collocation method commonly used. The disturbances are axisymmetrical jet flows normal to the mean flow and are similar to that used in [4].

$$\psi = \frac{1}{2}\varepsilon f(y) r^2 e^{-(r/l)^2} \tag{1}$$

$$(u, v, w) = (-\frac{x}{r^2}\psi_y, \frac{1}{r}\psi_r, -\frac{z}{r^2}\psi_y) \tag{2}$$

$$f(y) = (1 + y)^p (1 - y)^q \tag{3}$$

Here (x, y, z) and (u, v, w) are the streamwise, normal and spanwise coordinates and velocities, respectively, and $r^2 = x^2 + z^2$. The initial perturbations were specified by setting $p = 2, q = 5$ and $l = 1$ with various ε. This perturbation is stronger near the $y = -1$ wall than that near the other wall, and thus the perturbation is asymmetrical with respect to the center of the channel. In this work, the length scales, velocities and time have been normalized by the channel half height h, centreline velocity U_{cl} of the laminar parabolic velocity and h/U_{cl}, respectively. The simulations were conducted by using periodical boundary conditions in the streamwise and spanwise directions with $L_x = 16\pi$ and $L_z = 8\pi$, and for a non-dimensional time up to $T = 50$. The number of nodes used were $256 \times 97 \times 256$ and the skew-symmetrical form of the nonlinear terms was used to minimize the aliasing errors.

3 Results

Simulations were conducted for $Re = 1{,}000 - 6{,}000$ with $\varepsilon = 0.02$ and for $\varepsilon = 0.002 - 0.04$ with $Re = 3{,}000$. It is found that, for $\varepsilon = 0.02$, the perturbation energy $e = 1/2(< u^2 > + < v^2 > + < w^2 >)$ increases with time at a faster rate as Re increases. Here $< f^2 >= \int_0^{L_x} \int_0^{L_z} \int_{-1}^{1} f^2 dxddzdy$ at a given time is the total variance of f. The simulation time was not long enough to see the decay of the energy, but in general $< v^2 >$ decreases with time. For $\varepsilon = 0.02$ and $Re = 6{,}000$, the vertical variance $< v^2 >$ shows a dramatic increase with time after an initial decrease. This is an indication of the start of the transition. For $Re = 3{,}000$, it is found that perturbation energy increases faster with increasing ε, and at $\varepsilon = 0.04$ the vertical variance shows no sign of decrease after $T = 5$ as shown in Fig. 1b.

Figure 1 shows the development of e and $< v^2 >$ for the linear and full responses to the initial perturbation $\varepsilon = 0.04$ at $Re = 3{,}000$. The linear response was simulated by setting the nonlinear terms to zero (after the N-S equations were linearized around the laminar velocity profile) and is almost identical to that from the full response for the very small initial perturbation $\varepsilon = 0.002$. Figure 1a shows that the development of the perturbation energy e from the full response closely follows that from the linear response for $T < 10$ and the departure from the linear response for $T > 10$ indicates the effect from non-linearity. On the other hand, Fig. 1b shows that the nonlinear effect becomes important for $< v^2 >$ from $T \approx 5$, much early than that for e. In [4], it was also noticed that the signs of nonlinearity were first seen in the normal velocity and later in the other velocities through the forcing by the normal component. In the literature, especially in theoretical analysis, it has been a common practice to use the growth of perturbation energy as an indicator for the appropriateness of linear approximation. The results in Fig. 1 show that, in analysing the instability and transition of wall shear flows, the development of the vertical velocity may be a more appropriate indicator for identifying the start of the nonlinear effect.

One possible reason for this early nonlinear effect for the vertical velocity could be due to the large magnitude difference in the streamwise and vertical velocities. Figure 1 shows that, after the initial development, the magnitude of the streamwise velocity variance (the contribution to e is mainly from $< u^2 >$) is one order of

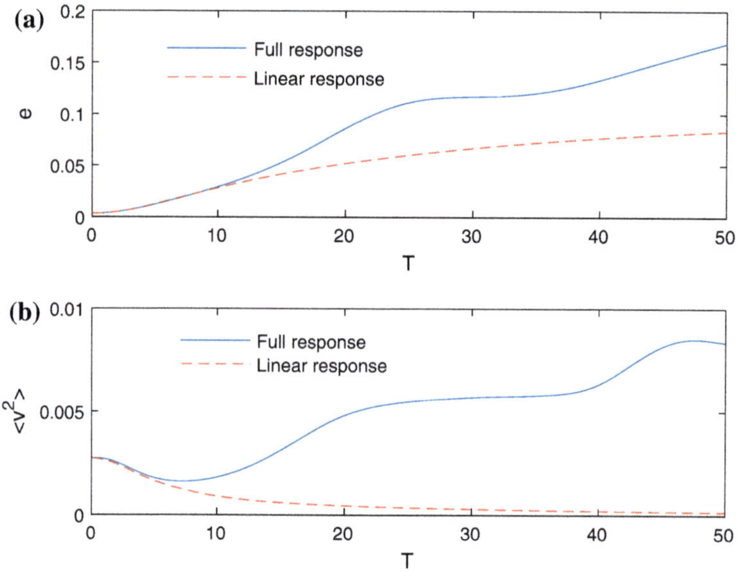

Fig. 1 Development of energy (**a**) and variance of vertical velocity (**b**) for linear responses and full responses including the nonlinearity. $Re = 3{,}000$ and $\varepsilon = 0.04$

magnitude higher than that of the vertical velocity, and it is expected that its contribution to the nonlinear terms in the vertical momentum equation will be large.

The spatiotemporal development of the flow structures for $Re = 3{,}000$ and $\varepsilon = 0.04$ was followed by creating movies based on velocity contours and isosurfaces based on vortex identification criteria $Q = 0.1$. It is found that the local disturbance generates a long and strong low speed streak among many weaker ones, and a pair of long streamwise vortices next to the strong low speed streak. This pair of vortices first appears at $T \approx 3.2$ almost vertically and becomes gradually stretched and elongated in the streamwise direction as their heads close to the centre of the channel moving downstream faster than their tails near the solid wall. The two streamwise vortices are close to each other and away from the wall downstream, further apart and close to the wall upstream, thus showing a small inclination to the wall. The first hairpin vortex appears above and at the middle of the strong low speed streak at $T \approx 31$ (this time depends on the Q value used for the current simulation). This is the primary vortex (the third hairpin vortex from the right in Fig. 2, between $30 < x < 32$) as identified in [7]. Several secondary hairpin vortices then appear sequentially behind and ahead of the primary hairpin vortex, and thus forming a packet of hairpin vortices as shown in Fig. 2 at $T = 45$. This packet is similar to that found in [7] which were simulated from a fully turbulent channel flow with initial perturbations based on linear stochastic estimation of extreme Reynolds shear stress events within the second quadrant. This shows that hairpin vortex packet can be formed in turbulent flows and in transition.

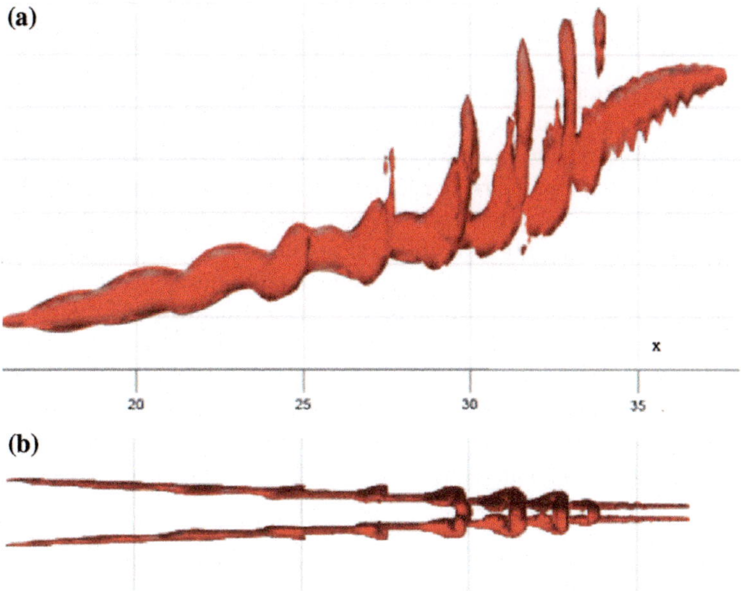

Fig. 2 Packet of hairpin vortices at $T = 45$, $Re = 3{,}000$ and $\varepsilon = 0.04$: **a** *side view*; **b** *top view*. The *vertical scale* has been stretched by a factor of 8

However, it is found that the "tongues" found in [7] are a natural part of the long streamwise vortex pair accompanied the low speed streak (as can be seen from the downstream part of Fig. 2b), rather than being generated after the appearance of the primary hairpin vortex. It is also found that the hairpin vortices are probably generated by the Kevin-Helmhotz type instability of the strong low speed streak, rather than by pinching-off the initial streamwise vortex due to the self induction of the Ω shaped hairpin vortex ahead. This conclusion was based on the observations that each individual hairpin vortex was first formed by a rolling up of the shear layer on top of the low speed streak and the waviness of the initial streamwise vortex pair as can be seen from Fig. 2a (especially that between $20 < x < 25$). The induction of the primary hairpin vortex also cannot explain the two hairpin vortices generated at its downstream (the two vortices from the right).

It is also found, from Fig. 2b, that for the hairpin vortex behind the primary one, small hairpin vortices were formed first near each leg of the initial streamwise vortex and paired together as they grow bigger while being convected downstream. Only the two hairpin vortices ahead of the primary one were formed directly without undergoing the pairing process. Inspection shows that the strong low speed streak is not a simple long blob of low speed fluid, rather it has two separate legs next to the two streamwise vortices, the two legs join together downstream. This again suggests that the formation of hairpin vortices is strongly associated with the local strong shear layer around the low speed streaks.

4 Discussion and Conclusions

The present simulations were conducted using symmetrical initial conditions with respect to $z = 0$ and periodic boundary conditions in the spanwise direction. Because of these, the packet of hairpin vortices are symmetrical. In real flows, perturbations will in general be non-symmetrical and boundary conditions non-periodical. These will result in asymmetrical structures. However, it is believed that the physical mechanisms for the formation of the hairpin vortices (unsymmetrical ones) are similar to those from the current simulations. The strong similarities between the packet of hairpin vortices from the current simulations using laminar base flow and that in [7] based on mean turbulent velocity profile shows that the formation of these structures may be a common feature for transition and fully developed turbulence in wall shear flows. However, it should also be noted that the wavepackets in the present simulations and those of [7] are long lasting and this could be because clean backgrounds have been used in both simulations. With a fluctuating turbulent background, the results in [9] show that hairpin vortices exist but are short lived.

Acknowledgements The author has benefited greatly from the discussions on instability and transition with Prof. C.B. Lee of Peking University, China.

References

1. M.V. Morkovin, The many faces of transition, in *Viscous Drag Reduction*, ed. by C. Wells (Plenum, Berlin, 1969)
2. N.D. Sandham, L. Kleiser, The late stages of transition to turbulence in channel flow. J. Fluid Mech. **245**, 319–348 (1992)
3. B.G.B. Klingmann, On transition due to three-dimensional disturbances in plane Poisenuille flow. J. Fluid Mech. **248**, 167–195 (1992)
4. D.S. Henningson, A. Lundbladh, A.V. Johansson, A mechanism for bypass transition from localized disturbances in wall-bounded shear flows. J. Fluid Mech. **250**, 169–207 (1993)
5. M.T. Landahl, Wave breakdown and turbulence. SIAM J. Appl. Math. **28**, 735–756 (1975)
6. T. Herbert, Secondary instability of boundary layers. Ann. Rev. Fluid Mech. **20**, 487–526 (1988)
7. J. Zhou, R.J. Adrain, S. Balachandar, T.M. Kendall, Mechanism for generating coherent packets of hairpin vortices in channel flow. J. Fluid Mech. **387**, 353–396 (1999)
8. J. Kim, P. Moin, R. Moser, Turbulent statistics in fully developed channel flow. J. Fluid Mech. **177**, 133–166 (1987)
9. G. Eitel-Amor, R. Orlu, P. Schlatter, O. Flores, Hairpin vortices in turbulent boundary layers. Phys. Fluids **27**, 025108 (2015)

Very Large-Scale Feature of Transitional and Turbulent Channel Flows: Dependence on Facilities

Shun Horii, Yoshiyuki Sagawa, Makoto Miyazaki and Masaharu Matsubara

Abstract A recent investigation revealed that a low frequency peak in premultiplied energy spectra due to turbulent patch passes in a transitional channel flow continues in a low Reynolds number but turbulent flow. In this study, to denial that a peculiar vibration of the facilities causes this low frequency peak, hot-wire measurements were performed. Spectral energy distributions obtained in different air channel flow facilities clearly confirm very large-scale feature at low Reynolds number and its independence from peculiarities of the facilities. A flow visualization experiment conducted in a water channel illustrates that a cluster of characteristic streaks forms into a inclined stripe whose shape is similar to a turbulent patch in a transitional channel flow. The streamwise and spanwise sizes of the cluster are in agreement with length scales estimated from the energy spectral and a spanwise correlation of the streamwise velocity. The similarity to the turbulent patch and continuity of the low-frequency peak from the transitional Reynolds number infer that the very large-scale feature relates to the intermittency in the transitional flow in terms of its maintenance and occurrence mechanism.

1 Introduction

Turbulent flows at high Reynolds number contain vortical structures of wide-range scales that extend over several orders of magnitude. While the smallest scale has been surveyed for half a century, it has been simply believed that geometry of a flow limits the largest scale like stirred water in a glass. It is questionable which dimension of flow geometry, whose length, width and height are very different such as pipe or channel flows, restricts the largest scale of flow.

It is well known that there exist large-scale features in boundary layer and pipe and channel flows at high Reynolds number. Kim and Adrian [1] observed a low-frequency plateau in spectral distribution of the streamwise velocity fluctuation

S. Horii · Y. Sagawa · M. Miyazaki · M. Matsubara (✉)
Mechanical Systems Engineering, Shinshu University, 4-17-1 Wakasato, Nagano 380-8553, Japan
e-mail: mmatsu@shinshu-u.ac.jp

R. Örlü et al. (eds.), *Progress in Turbulence VII*, Springer Proceedings in Physics 196, DOI 10.1007/978-3-319-57934-4_27

obtained by hot-wire measurements in a pipe flow. The length scale corresponding to the low frequency is 14 pipe radii, so then it was named very large-scale motions (VLSMs). Rake hot-wire measurements done by Monty et al. [2] in pipe and channel flows revealed the low frequency features have structure of meandering low-velocity streaks whose length scale is 25 radii or half channel widths. Hutchins and Marusic [3] performed rake-wire and single-wire experiments in a boundary layer and the velocity spectra demonstrates that the contribution of the superstructures, whose streamwise scale is 6 δ (δ is a boundary layer thickness), to the turbulence intensities increases with Reynolds number. The result with the rake-wire probe shows very long meandering features that have 20 δ streamwise length. They claimed that this meandering makes the peak length of the spectra smaller than the scale directly obtained by rake-wire or PIV experiments. Monty et al. [4], however, asserted that from velocity spectral distributions mapped in the wall-normal direction there exist VLSMs only in channel and pipe flows, not in a boundary layer.

In most of these investigations, the low frequency peak becomes unclear with decrease of Reynolds number, so that it is deemed that the large-scale features is regarded as a characteristic phenomenon at very high Reynolds number. Against this presumption, Seki and Matsubara [5] made hot-wire measurements for transitional and turbulent channel flow at low Reynolds number. In a transitional flow, they found a low-frequency peak in premultiplied energy spectra, which is due to turbulent patches passing through. It is surprising that the low-frequency peak which corresponds to streamwise length 25 d still exists even in a fully turbulent state, at $Re = 2660$. Reynolds number is defined as $Re = U_b d/\nu$, where U_b is bulk velocity, d is channel width and ν is kinematic viscosity. An extended investigation [6] confirmed that a low frequency peak in premultiplied energy spectra continues up to $Re = 4000$. This fact suggests strong relation between turbulent patches in a transitional flow and very large-scale features in a turbulent flow. However, one can doubt that the spectral peak and plateau are from specific characters of the experimental facility, such as mean velocity fluctuation induced by a blower. In order to clear up the doubt, hot-wire experiments in two different air channel facilities has been conducted. Furthermore, the spanwise correlation measurements and flow visualization were performed for determination of spanwise length and structure of the large-scale feature.

2 Experimental Setup

Small and large channel facilities were employed for the hot wire measurements. The small one is the same that Seki et al. [5] and Matsubara et al. [6] used. The wall distance d is 5.1 mm. Two tripping rods of 0.5 mm diameter are mounted on the entrance section walls 100 mm downstream from the outlet of the nozzle, so that the flow becomes fully developed turbulence 1000 mm downstream at the end of the entrance section. In the expansion section, the distance between the channel end walls is gradually widened from 190 to 260 mm in the 1000 mm length, so that the channel Reynolds number Re drops to 73%. This Re drop was for investigation of the relaminarizing channel flow and it is unnecessary for experiments on a turbulent flow. The measurements are performed at a 50 d downstream position from the outlet of the expansion section of 2800 mm length.

The other channel facility is much larger; the channel length and width are 17.35 m and 25 mm, respectively. The aspect ratio is 40. Two 2 mm diameter rods are fixed 440 mm downstream from the nozzle, and they trip the flow to turbulence. Realizing the seamless channel walls, the main part of the channel is made of 0.35 mm-thickness stainless plates of 16.9 m length and 1 m width. The aluminum angle members support the two plates for keeping them flat. The hot-wire measurements are made at the 130 mm upstream position from the channel outlet.

A constant-temperature hot-wire anemometer is used for the streamwise velocity measurements. Its sensor is made of a platinum wire of 2.5 μm diameter and 1 mm length. An analog/digital converter acquires the voltage from the hot wire anemometer with a sampling frequency of 20 kHz. The sampling time is 300 s for velocity spectral measurements, and 180 s for the correlation experiments. For the spanwise correlation measurements, two hot wires are independently mounted on moving devices, and the closest distance in the measurements is measured by a ruler to an accuracy of 0.5 mm.

For flow visualization experiments, a recirculated water channel facility is used. This facility has 8 m total streamwise length. The channel walls are made of glass and the channel width is 7.1 mm. A 0.9 mm diameter rod is mounted on one of the walls as a tripping wire at 220 mm downstream from entrance section. The spanwise distance is widened from 360 to 580 mm in expansion section so that Reynolds number is dropped to 62%. The aspect ratio is 82 in the 4550 mm test section. Flakes of pearl particles are used as tracer. The photographs are taken by a digital camera 4200 mm downstream from the outlet of the expansion section. In this study, x is streamwise direction, y is wall-normal direction and z is spanwise direction.

3 Result and Discussion

Figure 1 shows the premultiplied spectral of the streamwise velocity fluctuation at $Re = 2800$. The dimensional frequency f and the power spectral density Φ are normalized with d and U_b as $f^* = fd/U_b$ and $\Phi^* = \Phi d/U_b$. The inverse number of f^* corresponds to a length scale in unit of d. Although the high frequency peak moves to higher frequency with distance from the wall, the low frequency bump around $0.03 < f^* < 0.05$ almost stays at the same frequency for $y/d = 0.3$ and 0.5. The equivalent streamwise length scale of these bumps are from 20 to 30 d. Except small peaks around $f^* = 0.01$ for the large channel, the spectra distributions of the small and large channels are in good agreement, indicating that the very large-scale feature is an inherent phenomenon in a turbulent channel flow. The peaks around $f^* = 0.01$ might be from a peculiarity of the large channel facility. Two-point correlation was measured in the small channel in order to investigate the spanwise scale of the very large-scale feature. Dash lines shown in Fig. 2 are the correlation

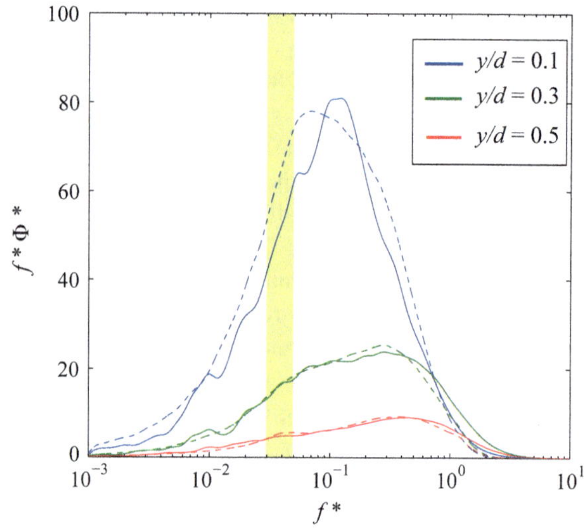

Fig. 1 The premultiplied spectral diagram of the streamwise velocity fluctuation measured in small and large facilities. The *dash lines* and the *solid lines* are results from the small and large channels, respectively. The frequency region in $0.03 < f^* < 0.05$ is highlighted

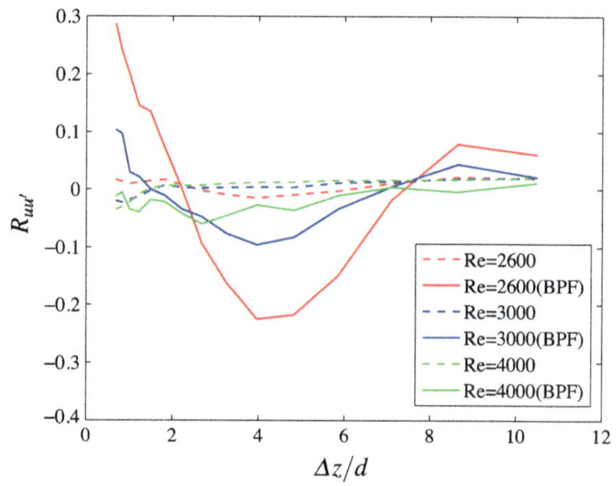

Fig. 2 Spanwise correlation of the streamwise fluctuation at center of the small channel

coefficient of the streamwise velocity fluctuation, $R_{uu'}$, and Δz is distance between two hot-wire probes. Although there is a little variation, they are close to zero except the negative correlation around $\Delta z/d = 1$. To extract contribution of the large-scale feature, band-pass filtering (BPF) for $0.03 < f^* < 0.05$ was employed. At $Re = 2600$, the correlation coefficient after the BPF begins from higher positive value, then it takes the minimum around $\Delta z/d = 4$. The minimum value is much lower than the unfiltered correlation. Assuming that this minimum corresponds to half spanwise wavelength of the very large-scale feature, it is estimated at eight channel widths. This spanwise length scale is also estimated at $Re = 3000$. For $Re = 4000$, the coefficient has the minimum around $\Delta z/d = 3$.

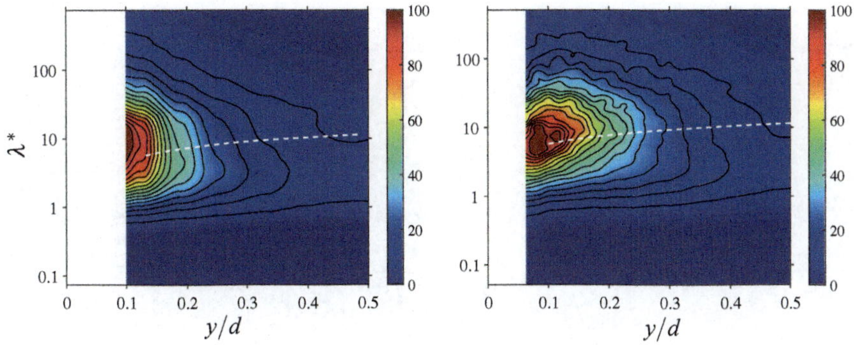

(a) Small channel.
(b) Large channel.

Fig. 3 Contour maps of streamwise velocity premultiplied spectra density at $Re = 2800$. The color variation represents the spectra density. The *white lines* are of (1)

Figure 3 shows contour maps of premultiplied spectra of the streamwise velocity fluctuation measured in both facilities at different distance from the wall, y. The non-dimensional wavelength λ^* is estimated from the frequency f as $\lambda^* = U_m/(f\,d)$, where U_m is a local mean velocity. Both contour maps have the highest peaks near the wall that indicates turbulent motion is the most energetic there. There exist two ridges of long and short wavelength, and the very large-scale feature still presents even far from the wall. The scale of the long wavelength is $\lambda \approx 10$–$20\,d$, which is longer than the wavelength of superstructures in boundary layer, $\lambda \approx 6\,\delta$ asserted by Hutchins and Marusic [3]. The white dash lines represent the peak position of Very Large-Scale Motion (VLSM) estimated by Monty et al. [4] defined as

$$\lambda^* = \frac{23}{2}\left(\frac{2y}{d}\right)^{3/7}. \tag{1}$$

Ridges, or plateaus, of the long wavelength in both facilities are slightly larger than the VLSM line.

Results of flow visualization conducted in a water channel facility to observe shape of the very large-scale feature are shown in Fig. 4. The flow direction is from left to right. In Fig. 4a, at $Re = 1500$, the turbulent patches indicate that the flow is in a transitional state. The turbulent patches are oblique and the streamwise length is several times larger than the channel width. Characteristic streak structures are also observed downstream and on sides of the patches. In Fig. 4b at $Re = 2600$ the flow is filled of turbulence. It was confirmed by Seki and Matsubara [5] from the analysis of velocity fluctuation data that this Reynolds number is the lowest for fully turbulent flow. Even in the fully turbulent state, a characteristic oblique region marked in a red circle is observed. This region contains streak structures whose streamwise and spanwise scales are slightly larger than those of the turbulent structure in other regions. Approximately, the streamwise spanwise sizes of the region are 20 and 10 d, respectively. These sizes correspond with the length scales estimated by the energy

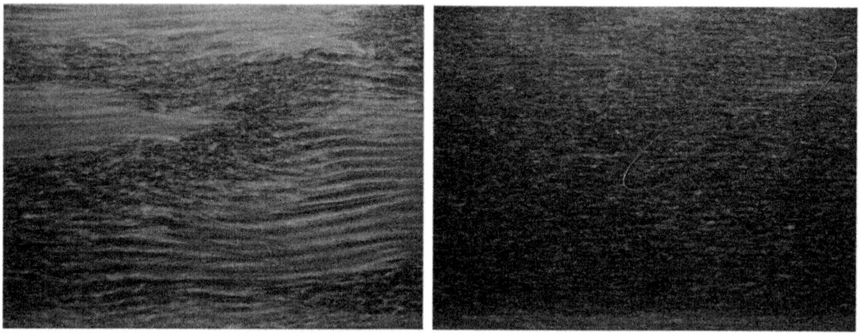

(a) $Re = 1500$. (b) $Re = 2600$.

Fig. 4 Flow visualization in water channel. The frame size is 31 $d \times 41$ d. In **b**, very large-scale feature is circled in *red*

spectra and the spanwise correlation in the air channel facilities and they are the same order of magnitude of those of the turbulent patches in a transitional channel flow as seen in Fig. 4a. Furthermore, Matsubara et al. [6] found that the low-frequency spectral peak due to turbulent patch continues as a bump even in fully turbulent states. These facts imply that the very large-scale feature in turbulent flow relates to the turbulent patch in the transitional flow in terms of maintenance and occurrence mechanisms.

4 Conclusions

The low-frequency plateau in the energy spectra is confirmed in a turbulent channel flow at low Reynolds number independently of the experimental facilities. The energy spectra and the spanwise correlation indicate that the very large-scale feature has 20–30 d streamwise and 8 d spanwise scales, which are similar size of the characteristic oblique region containing the large streak structures observed in the flow visualization. These results infer that the turbulent patch in a transitional channel flow is relevant to the very large-scale feature in turbulent flow with regard to their maintenance and occurrence mechanisms.

References

1. K.C. Kim, R.J. Adrian, Very large-scale motion in the outer layer. Phys. Fluids **11**(2), 417–422 (1999)
2. J.P. Monty, J.A. Stewart, R.C. Williams, M.S. Chong, Large-scale features in turbulent pipe and channel flows. J. Fluid Mech. **589**, 147–156 (2007)
3. N. Hutchins, I. Marusic, Evidence of very long meandering features in the logarithmic region of turbulent boundary layers. J. Fluid Mech. **579**, 1–28 (2007)

4. J.P. Monty, N. Hutchins, H.C.H. Ng, I. Marusic, M.S. Chong, A comparison of turbulent pipe, channel and boundary layer flows. J. Fluid Mech. **632**, 431–442 (2009)
5. D. Seki, M. Matsubara, Experimental investigation of relaminarizing and transitional channel flows. Phys. Fluids **24**, 124102 (2012)
6. M. Matsubara, S. Horii, Y. Sagawa, Y. Takahashi, D. Saito, Very-large-scale fluctuations in turbulent channel flow at low Reynolds number. Int. J. Heat Fluid Flow **62B**, 593–597 (2013)

Part VI
Miscellaneous Topics

Turbulence Structure Analysis of DNS Data Using DMD and SPOD: Mixing Jet and Channel Flow

Abouelmagd Abdelsamie, Gábor Janiga, Cheng Chi and Dominique Thévenin

Abstract The objectives of this work are to analyze and investigate the turbulence structures in a channel flow and in a mixing jet using Dynamic Mode Decomposition (DMD) and Snapshot Proper Orthogonal Decomposition (SPOD). The analyzed data have been generated by Direct Numerical Simulation at high Reynolds numbers. In the channel flow, the occurrence of turbulent superstructures is mainly examined. The jet case is employed to investigate mixing in a turbulent jet flow. In both cases, DMD and SPOD analysis are compared to test their performance concerning the analysis of complex flows and to highlight the complementarity between these two approaches.

1 Fundamentals of SPOD and DMD

In principle, SPOD and DMD are designed to defined the coherent structure of turbulence but in two different ways. In the following, the basic equations for each of them are defined.

In **SPOD**, each signal (here, flow velocity) can be decomposed into temporal a and spatial parts ϕ,

$$\mathbf{u}(\mathbf{x}_i, t_j) = \mathbf{u}_j = \sum_{m=1}^{M} a_m(t_j)\phi_m(\mathbf{x}_i), \tag{1}$$

A. Abdelsamie (✉) · G. Janiga · C. Chi · D. Thévenin
University of Magdeburg "Otto von Guericke", Universitätsplatz 2,
39106 Magdeburg, Germany
e-mail: abouelmagd.abdelsamie@ovgu.de

G. Janiga
e-mail: janiga@ovgu.de

C. Chi
e-mail: cheng.chi@ovgu.de

D. Thévenin
e-mail: thevenin@ovgu.de

R. Örlü et al. (eds.), *Progress in Turbulence VII*, Springer Proceedings
in Physics 196, DOI 10.1007/978-3-319-57934-4_28

where the eigenfunctions ϕ_m are called spatial modes, and a_m are the temporal coefficient. The ϕ_m and a_m are completely determined after solving the eigenvalue problem and obtaining the corresponding eigenvalues λ, and eigenvectors [2].

The basics of **DMD** is to assume that the snapshots are generated by a linear, discrete time model,

$$\mathbf{u}_{j+1} = \mathbf{A}\mathbf{u}_j, \tag{2}$$

It is assumed that the snapshots become linearly dependent for an increasing number of snapshots, so that snapshot $\mathbf{u}_N$ can be constructed by a linear combination of all previous snapshots. Following [7], an eigenvalue problem can be derived. With computing the eigenvalues $\tilde{\lambda}$ and eigenvectors $\tilde{\mathbf{v}}_i$, the DMD modes and temporal amplitudes can be obtained [4]. In DMD, the eigenvalues and eigenvectors are complex numbers. The frequency f_i and growth (or decay) rate σ_i of DMD are obtained by logarithmic mapping of the eigenvalues; $f_i = \Im[\ln(\tilde{\lambda}_i)]/(2\pi\Delta t)$, and $\sigma_i = \Re[\ln(\tilde{\lambda}_i)]/\Delta t$.

The main difference between SPOD and DMD is that SPOD is based on an energy ranking of orthogonal structures computed from a correlation matrix of the snapshot. This leads to two possible drawbacks: (1) The energy content is not necessarily the correct measure to rank the flow structures; (2) Valuable information may be lost since only second-order statistics are used as a basis for the decomposition. Additionally, POD allows mixing between scales and frequency at each mode; while DMD separately provides one growth rate and frequency for each mode. In DMD, the dynamic modes are non-orthogonal; they can be sorted by frequency, growth rate, or mode norm. SPOD and DMD also differ in the computational requirements. For the following examples, it has been found that computing the DMD decomposition is six times slower than for POD.

2 Numerical Settings

Direct Numerical Simulations (DNS) have been conducted to simulate two different cases: (1) turbulent flow in a square-shaped channel with bulk Reynolds number of 20,000 in a domain of size $H \times 20H \times H$, where H is the height of the channel; this domain was discretized with about 400 million grid points. (2) mixing of hydrogen with air in a turbulent jet flow at a jet Reynolds number of 2 600, simulated with about 70 million grid points in a domain of size $7.5d_j \times 30d_j \times 7.5d_j$, where d_j is the jet diameter. All simulations have been performed with the in-house DNS code “DINO” [1]. The results of these DNS have been then analyzed with the in-house Python script called PyPODe [2]. PyPODe has been coded on top of the modred-1.0.2 package [6] and contains both DMD [7] and SPOD [8] algorithms. Figure 1, top, bottom show exemplary 2D-planes of velocity magnitude in the turbulent channel and the iso-volume of mixture fraction in the jet, respectively.

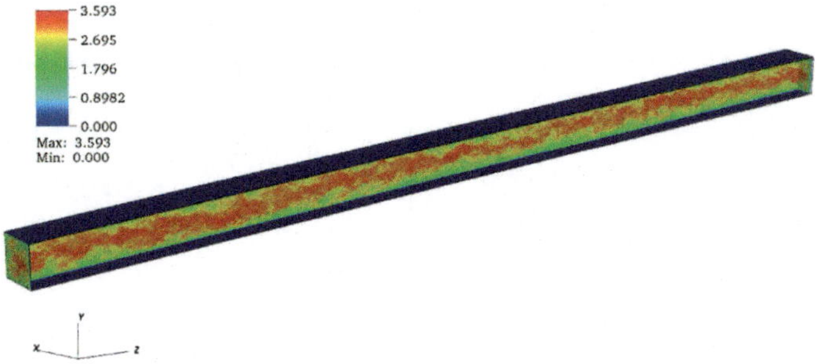

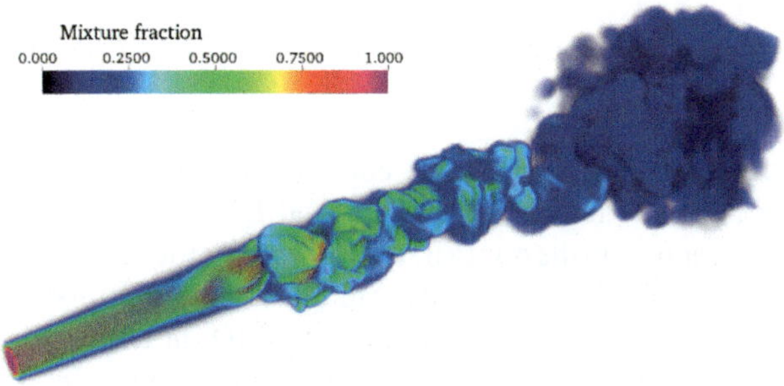

Fig. 1 *Top*: 2D cuts showing instantaneous velocity magnitude in the turbulent channel flow. *Bottom*: Iso-volume of instantaneous mixture fraction in the turbulent jet

3 Results and Analysis

In this section, the analysis of jet flow and turbulent channel flow using both of POD and DMD will be discussed. In the following, the first mode (which represents the mean value) is removed and not included in the spectrum or to show the spatial modes.

Jet flow As in many turbulent mixing process, the turbulent jet has three different ranges of scales responsible for the entire turbulence dynamics [3]: (1) Smallest scales that are responsible for molecular mixing; (2) Large scale flow structures describe the entrainment stage that is responsible for the engulfment of large pockets of irrotational fluid species into the turbulent flow region; (3) Intermediate range of scales, which is responsible for the subsequent kinematic stirring process responsible for the large interfacial surface generation between the mixing species. Figure 2, left shows the eigenvalue (spectrum) of SPOD versus number of modes. From this figure it is obvious that the first modes (most energetic modes) are the most dominant, with a rapid energy decrease (exponential decay). By comparison, the logarithmic mapping of the DMD eigenvalues can be represented as growth (decay) rate versus frequency, as seen in Fig. 2, right. In this figure, the numbers represent the numbering of the modes and are not ordered; one could sort them based on the frequency in order to identify the most dominant modes. Figure 3 shows the iso-surfaces of

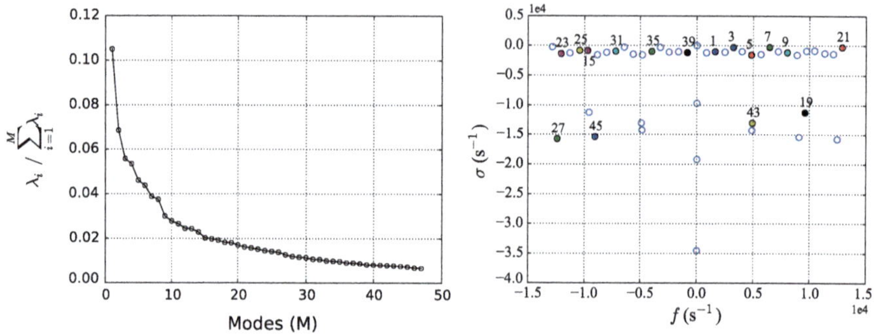

Fig. 2 Modal spectrum of the mixing jet case. *Left*: eigenvalue of SPOD. *Right*: spectrum of DMD

the spatial modes: top row is the SPOD; bottom row is the DMD. From the first five modes (Fig. 3a–e and h–l) in both SPOD and DMD, it is clear that the most dominant structures appear in the original shear layer and progressively extend toward the transition region. The first four modes are quite similar in topology, while the fifth mode shows difference between SPOD and DMD; in DMD (Fig. 3l) broader structures appear at the head of the jet, which may have been lost in SPOD (Fig. 3e) due to its intrinsic limitation to second-order. Figure 3f, m illustrate that SPOD and DMD show qualitatively intermediate modes. At high frequency in DMD, Fig. 3n shows the fastest and smallest structure revealed by the dynamics modes. It still shows a similar structure to the largest mode in SPOD (Fig. 3g), even if it is known that this one can not be directly interpreted in terms of a physical structure.

As conclusion, the SPOD, which is computationally six times faster than DMD, is able to extract properly the most energetic modes and their corresponding coherent structures. In case the interest is set on low-order structures or on specific frequency regions, DMD will probably be the proper choice.

Channel flow In very long turbulent channel flow at high Reynolds numbers (Re $\geq$ 20 000), long meandering velocity fluctuations with both positive and negative streamwise velocity are sometimes observed [5]. These velocity fluctuations appear in the log-law and lower wake regions of the turbulent boundary layer and are named superstructures. In the present work, possible superstructures are tracked with SPOD and DMD.

As usual the spectrum of modal decomposition should be presented first (Fig. 4, left, right). Again, the first dominant spatial modes could be represented isolately based on the given spectrum. Figure 5 shows that SPOD (Fig. 5, left), and even more DMD (Fig. 5, right) reveal the presence of very long meandering velocity fluctuations, with both positive and negative streamwise velocity. These might be superstructures, in agreement with the definition of [5]. However, it is in this case difficult to find a qualitative similarity between the POD and DMD modes. This is attributed to the fact that the number of snapshots employed for the analysis is not sufficient. In our future work, a larger number of snapshots must be taken into account. From these

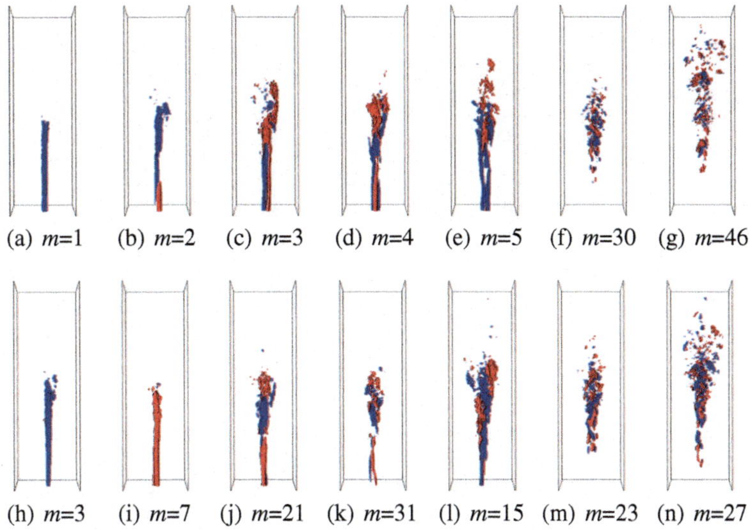

Fig. 3 Iso-surfaces of the spatial modes in the mixing jet, colored by the sign of streamwise velocity (positive or negative); each mode is normalized by its local maximum. *Red* and *blue* iso-surfaces represent the values of 0.5 and −0.5, respectively. *Top*: SPOD. *Bottom*: DMD

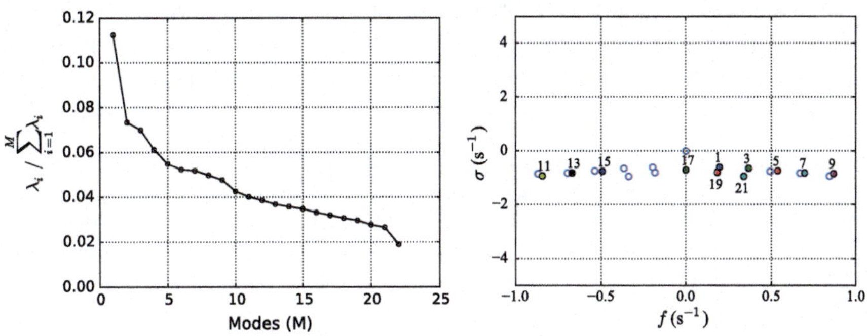

Fig. 4 Modal spectrum of the channel flow case. *Left*: eigenvalue of SPOD. *Right*: spectrum of DMD

preliminary results, it appears that DMD might play an essential role to follow the dynamics of superstructures; SPOD can probably only be used for very energetic superstructures.

4 Conclusion

Turbulent structures and mixing in a long turbulent channel flow and in a mixing jet have been investigated with both SPOD and DMD modal decomposition methods. It has been found that in the mixing jet case, both SPOD and DMD deliver

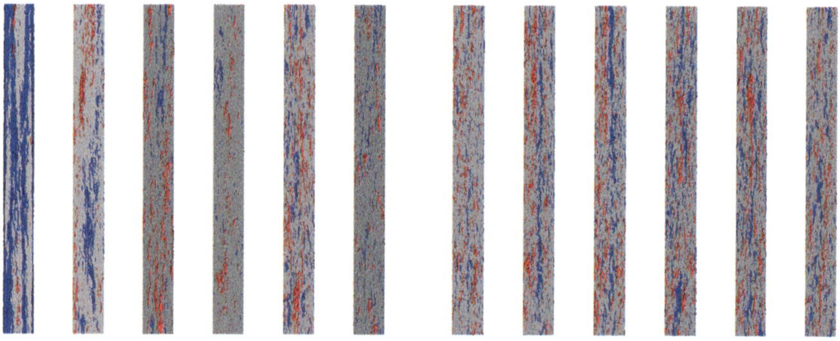

(a) $m=1$ (b) $m=2$ (c) $m=3$ (d) $m=4$ (e) $m=19$ (f) $m=21$ (g) $m=1$ (h) $m=3$ (i) $m=5$ (j) $m=7$ (k) $m=9$ (l) $m=11$

Fig. 5 Iso-surfaces of the spatial modes in the channel flow close to the wall (part of the domain), *colored* by the sign of streamwise velocity (positive or negative); each mode is normalized by its local maximum value. *Red* and *blue* iso-surfaces represent the values of 0.5 and −0.5, respectively. *Left* POD. *Right* DMD

qualitatively very similar information concerning dominant structures and modes. In the turbulent channel flow at high Reynolds number, DMD seems to be more suitable for identification and tracking of possible superstructures. However, it is found in both cases that SPOD computations are about six times faster than the corresponding DMD analysis. As a final conclusion, SPOD and DMD should probably be retained as complementary methods for the same analysis, taking advantage of both approaches.

Acknowledgements The financial support of the International Max Planck Research School Magdeburg for Advanced Methods in Process and Systems Engineering (IMPRS ProEng) is gratefully acknowledged. The computer resources provided by the Gauss Center for Supercomputing/ Leibniz Supercomputing Center Munich under grant pro84qo have been essential to obtain the DNS results presented in this work.

References

1. A. Abdelsamie, G. Fru, T. Oster, F. Dietzsch, G. Janiga, D. Thévenin, Towards direct numerical simulations of low-Mach number turbulent reacting and two-phase flows using immersed boundaries. Comput. Fluids **131**, 123–141 (2016)
2. P. Arányi, G. Janiga, K. Zähringer, D. Thévenin, Analysis of different POD methods for PIV-measurements in complex unsteady flows. Int. J. Heat Fluid Flow **43**, 204–211 (2013)
3. P.E. Dimotakis, The mixing transition in turbulent flows. J. Fluid Mech. **409**, 69–98 (2000)
4. O. Frederich, D.M. Luchtenburg, Modal analysis of complex turbulent flow, in *7th International Symposium on Turbulence and Shear Flow Phenomena (TSFP-7)*, 2011. Ottawa, Canada
5. N. Hutchins, I. Marusic, Evidence of very long meandering features in the logarithmic region of turbulent boundary layers. J. Fluid Mech. **579**, 1–28 (2007)
6. MODRED 2.0.1 package, http://pythonhosted.org/modred/index.html
7. P.J. Schmid, Dynamic mode decomposition of numerical and experimental data. J. Fluid Mech. **656**, 5–28 (2010)
8. L. Sirovich, Turbulence and the dynamics of coherent structures. Part 1: coherent structures. Q. Appl. Math. **45**(3), 561–571 (1987)

Signature of a Cubical Canopy on the Spatial Dynamics of an Atmospheric Boundary Layer

Jérémy Basley and Laurent Perret

Abstract The present work investigates the influence of the plan density of a cube-based canopy on the coherent structures developing in the roughness sub-layer of an atmospheric boundary layer. Stereo Particle Image Velocimetry is employed using a large field of view in a horizontal plane, so as to identify both the large meandering structures of the boundary layer and the dynamics associated with the wake of the canopy. Pre-multiplied two-dimensional spectra show wall-normal velocity fluctuations are constrained by the pitch of the canopy.

1 Context

Recent results obtained for high-Reynolds number boundary-layers over smooth-walls have shed light on the nature of the coupling between the near-wall turbulence and the large-scales of the flow developing above [6]. Concurrently, attention has been devoted to the structure of atmospheric boundary layer flows developing over urban or vegetation canopies, demonstrating similarities between smooth- and rough-wall-bounded flows. In particular, the presence of streaky patterns of low- and high-speed regions, of ejections and sweeps associated to the hairpin model and the organization of hairpin vortices in packets have been evidenced [4].

Due to the nature of the roughness elements, urban-like canopy flows are characterized by a strong multi-scale character both in space and time. This leads to a dramatic alteration of the near-wall cycle of turbulence, accompanied by complex inter-scale interactions between the canopy and the boundary layer flows [2]. Recent studies focus on the effect of canopy frontal or/and planar densities on the roughness sub-layer (see the extensive parametric study [7]). The direct numerical simulations (DNS) conducted in [3] have shown that the characteristic length scales within the

J. Basley (✉) · L. Perret
LHEEA, Ecole Centrale Nantes (CNRS UMR 6598), BP 92101,
44321 Nantes Cedex, France
e-mail: jeremy.basley@ec-nantes.fr

L. Perret
e-mail: laurent.perret@ec-nantes.fr

R. Örlü et al. (eds.), *Progress in Turbulence VII*, Springer Proceedings
in Physics 196, DOI 10.1007/978-3-319-57934-4_29

roughness sub-layer increase with the canopy height but only a single plan density was studied (25%). The case of a sparse cubical canopy (12.5% plan density) was treated by Lee an co-workers: low-speed streaks are drastically shortened due to the separation/wake at the tip of the cubes, while friction velocity remains basically constant with the periodicity of the canopy [5]. The present experimental work focuses specifically on the effect of canopy periodicity on the length scales and extend to higher Reynolds numbers not yet accessible to DNS.

2 Experiments

A turbulent boundary layer (thickness $\delta \sim 1$ m) develops over a 20 m long fetch of staggered cubes ($h = 50$ mm high) in an atmospheric wind-tunnel, yielding high Reynolds numbers $Re_\delta = u_* \delta/\nu = 25{,}000$ ($Re_h = 1250$). The external velocity is set to $U_e = 5.8$ m/s. There is a weak streamwise pressure gradient $dP/dx = -0.37$ Pa/m for the canopy coverage $\beta = 25\%$, with negligible impact on the boundary layer development. Details on the characterisation of the facility can be found in [2].

Three different plan densities, 6.25, 25 and 44.4%, respectively, are investigated to determine the influence of the canopy horizontal length scales on the dynamics of the roughness sub-layer. Figure 1 sketches the canopies along with the pitch of the patterns.

Measurements were conducted using three-component two-dimensional Stereo Particle Image Velocimetry (SPIV) in a horizontal plane at $z = 1.5h$ within the roughness sub-layer (Fig. 2). The region of interest extends over about $11h \times 10h$ (or $0.55\delta \times 0.5\delta$) with about 20×12 vectors/h^2 (32×32 pixel windows), to encompass both the largest scales of the flow and the obstacle-induced coherent structures. For each configuration, five series of 2100 velocity fields were acquired at $f_{spiv} = 7$ Hz. These 10,500 samples correspond to about 9000 turnover times (δ/U_e) while the sampling frequency offers a (minimal) overlap of two successive fields.

Constant-temperature hot-wire measurements were also performed downstream of the SPIV field using a 5 μm-diameter 1.25 mm-long hot-wire probe (DANTEC-

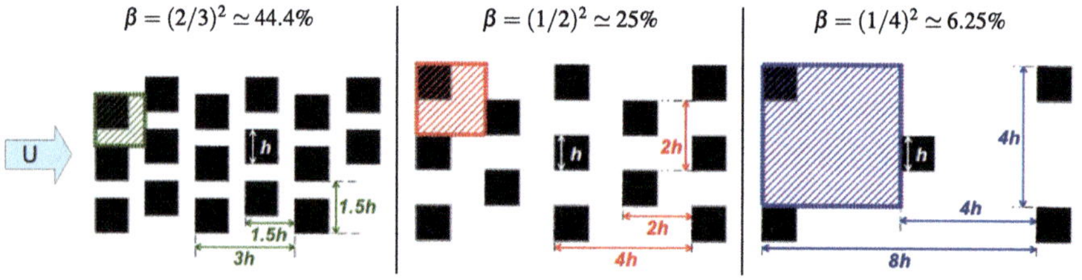

Fig. 1 Top view of the three canopy staggered distributions, with the flow coming from the left. The *hatched region* represents the periodic pattern for each case

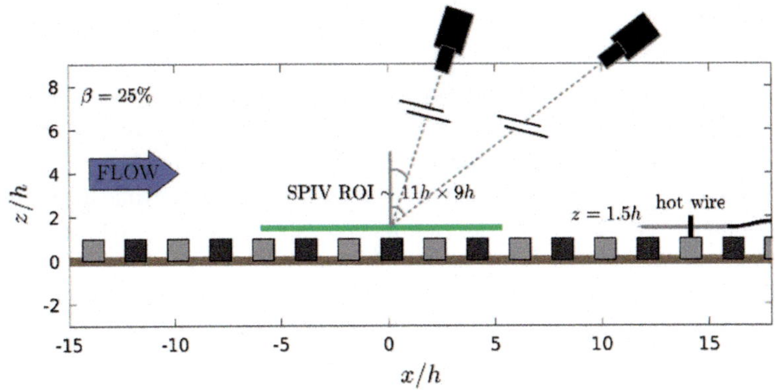

Fig. 2 Side-view of the experimental set-up for stereo-PIV hot-wire-synchronised measurements

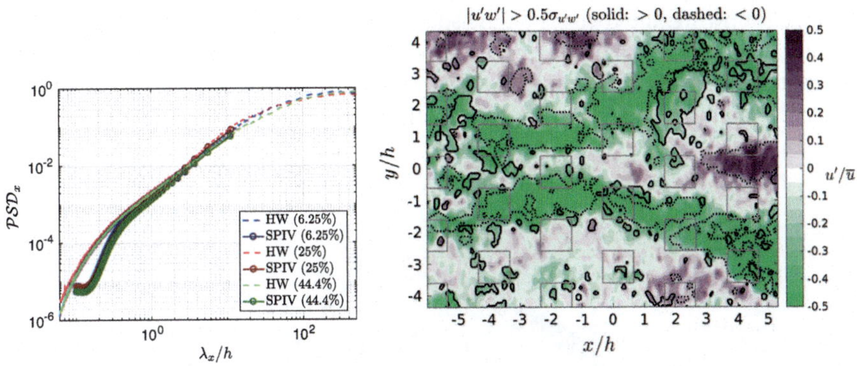

Fig. 3 *Left* power spectra E_u for all three canopy configurations. SPIV data are compared to synchronised hot-wire data using Taylor's hypothesis. *Right* SPIV snapshot for canopy $\beta = 25\%$. *Colors* represent fluctuating velocity u' normalised by mean flow $\overline{u}$, while *solid*, respectively *dashed*, *black lines* depict positive, resp. negative, *contours* of Reynolds wall-normal shear stress $u'w'$

55P11). Streamwise velocity time-series were acquired for an overall duration of 100 min, that is about 35,000 turnover times, at sampling frequency $f_{hw} = 10\,\text{kHz}$. SPIV and hot-wire data are compared using Taylor's hypothesis on hot-wire time series in Fig. 3, where resulting spectra match quite well for a range of $0.4 < \lambda_x/h < 11$. SPIV starts low-pass filtering velocity fluctuations below $\lambda_x \sim 0.4$. Each velocity field $\mathbf{u}_i(x, y)$—Fig. 3—is ensemble averaged ($\langle\cdot\rangle$) and decomposed as

$$\mathbf{u}_i(x, y) = \langle\mathbf{u}\rangle(x, y) + \mathbf{u}'_i(x, y) = (\langle u\rangle + u'_i)\mathbf{e_x} + (\langle v\rangle + v'_i)\mathbf{e_y} + (\langle w\rangle + w'_i)\mathbf{e_z}, \quad (1)$$

3 Analysis

The power spectra depicted in Fig. 3 show remarkable resemblance between the three canopy densities. This points out that the streamwise velocity component alone does not provide any specific features depending on the canopy density. Since u' dynamics is mainly governed by elongated super-structures stretching far beyond the roughness sub-layer, those very large structures are indeed expected to scale on the boundary layer thickness δ rather than any near-wall scales.

On the other hand, two features of the spatial dynamics can be strongly influenced by the canopy. First, the wall-normal velocity fluctuations w' represent the wall-normal exchanges between near wall and inertial regions of the boundary layer. In particular Q2 and Q4 quadrants of shear stress $u'w'$ are related to ejections and sweeps, respectively (Fig. 3). Second, the swirling stress λ_{ci}^{s} [1] may identify wake dynamics generated by the cubical canopy.

In the following, the focus is on the spatial dynamics of these two quantities. To that aim, two-dimensional power spectral densities are computed from the 10,500 snapshots and then ensemble averaged, for all three canopies. The 95% confidence intervals estimated using a normal distribution are within the thickness of the plot

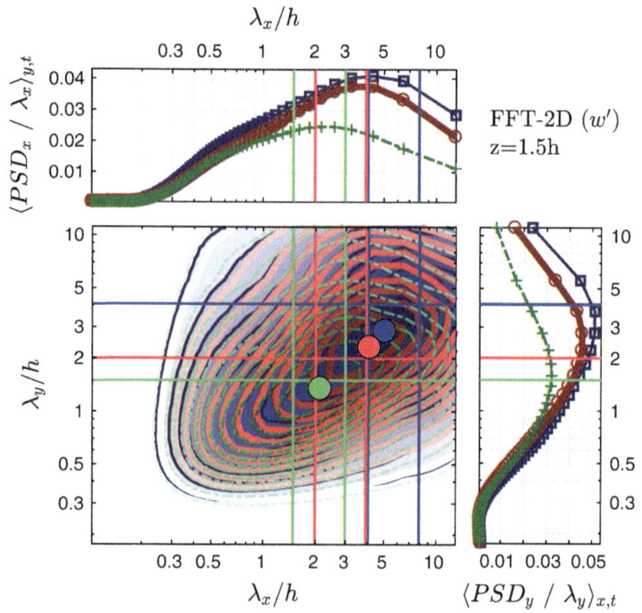

Fig. 4 Pre-multiplied two-dimensional spatial power spectral density (*PSD*) for wall-normal velocity fluctuation w'. *Top* spanwise-integrated streamwise PSD and *right*, streamwise-integrated spanwise PSD. The three canopy densities are shown: $\beta = 6.25\%$ in *blue* (*squares* and *filled contours*), $\beta = 25\%$ in *red* (*thick lines*) and $\beta = 44.4\%$ in *green* (*dashed lines*). *Colored circles* mark the energy peaks and *straight lines* highlight the pitch for the three canopies

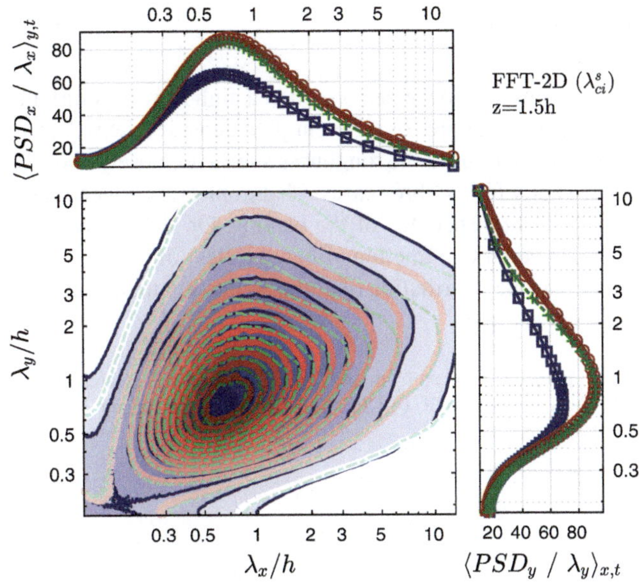

Fig. 5 Pre-multiplied two-dimensional spatial power spectral density (*PSD*) for swirling strength λ^s_{ci} signed by out-of-plane vorticity. See caption of Fig. 4 for details

lines. Note that only the length scales associated with the fluctuating (and advected) dynamics are investigated: the sample averaged field is removed prior to spectral analysis so that its spatial distribution does not influence the results.

Wall-normal exchanges are characterised by fluctuating velocity component w' through Reynolds shear stresses $u'w'$ and $v'w'$. The results are therefore similar for all three quantities. Figure 4 depicts the energy spectra of w'. It shows unambiguously that the energy peak is shifted depending on the periodicity of the canopy pattern, though the correspondence is less strict for the sparse canopy (6.25%). This suggests that the cube spacing induces a confinement effect on the length scales of the sweeps and ejections.

The eddy dynamics is investigated using the same spectral analysis applied to vorticity-signed swirling strength λ^s_{ci}. Results are plotted in Fig. 5. Contrarily to what was seen for w'-related quantities, λ^s_{ci} spectrum does not exhibit any significant change with the canopy coverage β: the peak of energy is located around $(\lambda_x, \lambda_y) = (0.6h, 0.7h)$ for the three canopies. Such a length scale is consistent with wake dynamics from the canopy. Furthermore, the spectrum associated with $\beta = 6.25\%$ is less energetic than the other two cases. This also points to wake-induced structures: a lower density of cubes corresponds to fewer wakes but does not change the length scales associated with shed eddies. Nevertheless more insight would be required to ascertain the nature of these eddies.

4 Concluding Remarks

Stereo PIV measurements have been performed in a high Reynolds turbulent boundary layer over three urban canopies of different plan densities. Length scales associated with w'-related fluctuations in the roughness sub-layer closely correspond to the canopy pattern periodicity for the three cases. That implies that sweeps and ejections are constrained by the cube spacings of the canopy. On the contrary eddy length scales remain constant equal to a fraction of h, regardless of the canopy periodicity. Ongoing work concerns the distribution of eddies relatively to the large meandering structures of the boundary layer.

Acknowledgements Financial support from the French National Research Agency through research Grant URBANTURB No. ANR-14-CE22-0012-01 is acknowledged.

References

1. R.J. Adrian, K.T. Christensen, Z.-C. Liu, Analysis and interpretation of instantaneous turbulent velocity fields. Exp. Fluids **29**, 275–290 (2000)
2. K. Blackman, L. Perret, Non-linear interactions in a boundary layer developing over an array of cubes using stochastic estimation. Phys. Fluids **28**, 095108 (2016)
3. O. Coceal, A. Dobre, T.G. Thomas, S.E. Belcher, Structure of turbulent flow over regular arrays of cubical roughness. J. Fluid Mech. **589**, 375–409 (2007)
4. A. Inagaki, M. Kanda, Organized structure of active turbulence over an array of cubes within the logarithmic layer of atmospheric flow. Bound. Layer Meteorol. **135**(2), 209–228 (2010)
5. J.H. Lee, A. Seena, S.-H. Lee, H.J. Sung, Turbulent boundary layers over rod- and cube-roughened walls. J. Turbul. **13**(40), 1–26 (2012)
6. I. Marusic, B.J. McKeon, P.A. Monkewitz, H.M. Nagib, A.J. Smits, K. Sreenivasan, Wall-bounded turbulent flows at high Reynolds numbers: recent advances and key issues. Phys. Fluids **22**(6), 065103 (2010)
7. M. Placidi, B. Ganapathisubramani, Effects of frontal and plan solidities on aerodynamic parameters and the roughness sublayer in turbulent boundary layers. J. Fluid Mech. **782**, 541–566 (2015)

Near and Far-Field Analysis of an Axisymmetric Fractal-Forced Turbulent Jet

Massimiliano Breda and Oliver R.H. Buxton

Abstract In this paper, the role of the initial conditions in affecting the flow physics in the near-field, and the evolution towards self-similarity, of an axisymmetric turbulent jet is examined. The near-field large scale coherent structures are manipulated with the aid of noncircular geometries, such as square and fractal exits. Planar PIV and hot-wire anemometry are deployed to study the flow both spatially and temporally. Despite the significant alteration of the near-field flow physics due to the different exit geometries, it is found that the evolution towards self-similarity is comparable between all jets. Moreover, non-equilibrium dissipation is found between 24 and 26 equivalent diameters D_e downstream of the jet exit where mean velocity and Reynolds stresses are self-similar, suggesting the microscales of the flow take much further than previously thought to regain the classical scaling laws.

1 Introduction

The aim of this paper is to investigate how the initial conditions affect the evolution of an axisymmetric jet towards a self-similar state. This topic has been previously analysed, however contrasting conclusions were reached. Antonia and Zhao [2] showed that axisymmetric jets with a fully developed and a top-hat mean velocity exit profile reached self-similarity after the same streamwise distance. George [3] stated that the initial conditions could affect growth and spreading of a jet, but not the self-similar scalings. In order to achieve a further insight in this topic, noncircular geometries are used to affect the near-field coherent structures. The latter are responsible for the development of an axisymmetric jet through the entrainment of background fluid into the turbulent stream. The rate of entrainment, which describes the exchange of momentum, energy and various properties between the irrotational surrounding flow and the jet stream, is strongly affected by large scale

M. Breda (✉) · O.R.H. Buxton
Imperial College London, London, UK
e-mail: mb2014@ic.ac.uk

O.R.H. Buxton
e-mail: o.buxton@imperial.ac.uk

R. Örlü et al. (eds.), *Progress in Turbulence VII*, Springer Proceedings
in Physics 196, DOI 10.1007/978-3-319-57934-4_30

(engulfment) and small scale processes (nibbling). Hence, noncircular geometries are used to affect the near-field development of the jet's large scale coherent structures, the so called Kelvin–Helmholtz (K–H) vortices, and to verify how different near-field entrainment rates and mixing properties affect the jet evolution towards the far-field. Previous studies on the effect of noncircular geometries on entrainment and mixing by Gutmark and Grinstein [5], amongst others, have observed that a combination of corners and large wetted perimeter was highly beneficial for combustion since it allowed combining an increased intensity of small-scale turbulence at the corners (and hence an improved small scale mixing), with the coherent structures at the flat-sides. Moreover, Shakouchi and Iriyama [9] observed that a large wetted perimeter, such as for their petal nozzle, was highly beneficial for mixing. Hence, the properties of a square and fractal exit, which allows maximising the number of corners and the wetted perimeter for a given open area, are compared to a round jet.

The same fractal geometry has been previously studied on axisymmetric wake generators. In the near-field, a break-up of the large scale coherent structures and a reduction of the shedding energy was observed by Nedić et al. [7]. As previously mentioned, the near-field of an axisymmetric jet also contains coherent structures; hence it will be verified if a similar breaking-up of these coherent structures occurs there. Further downstream, the initial conditions also proved to strongly affect the wake's evolution towards the self-similar state. Antonia and Pearson [1] verified that the mean dissipation rate of wakes was strongly dependent on the initial conditions. Nedić et al. [8] found non-equilibrium dissipation in the near-field and up to a distance significantly far away from the wake generator, where self-similar profiles were observed. Non-equilibrium dissipation, as detailed by Vassilicos [12], is found in regions where the inter-scale energy transfer is not instantaneously in equilibrium with the dissipation rate. The classical Kolmogorov theories require C_ε, as in (1), to be a constant, however when non-equilibrium dissipation is present C_ε is observed to be a function of Re_G and of the local Reynolds number $Re_\lambda(x)$ as in (2).

$$C_\varepsilon \equiv \frac{\varepsilon \mathscr{L}}{u'^3} \sim constant \tag{1}$$

$$C_\varepsilon \equiv \frac{\varepsilon \mathscr{L}}{u'^3} \sim \frac{Re_G^{1/2}}{Re_\lambda} \neq constant \qquad Re_\lambda \neq constant \tag{2}$$

Here, ε is the dissipation, $\mathscr{L}$ the integral length scale, λ the Taylor microscale, Re_λ the Taylor Reynolds number and u' the root-mean-square of the streamwise velocity fluctuations. Non-equilibrium dissipation was also found for square and fractal grids, as detailed by Hearst and Lavoie [6], where it was shown C_ε varied according to (2) for an extensive region of space, starting from very close to the grid until it transitioned back to the classical $C_\varepsilon \sim constant$. Despite the evidence of non-equilibrium dissipation combined with large scale self-similar profiles, in the self-similar regions of an axisymmetric jet the momentum flux is constant and hence Re_λ should not vary. Therefore it is not expected to find signs of non-equilibrium dissipation in the self-similar state, as in other shear flows. The first part of this paper will focus on assessing how the exit geometries effect the near-field coherent structures. It will be then followed by an analysis of entrainment and mixing. The entrainment rate will then be compared between the jets, calculating the entrainment parameter α as in [11].

$$\alpha = -\frac{\delta_m}{2\gamma_m} + \frac{Q}{2M^{1/2}}\frac{\partial}{\partial x}(\ln \gamma_m) \tag{3}$$

This equation takes into account various parameters, such as turbulence production (δ_m), energy flux (γ_m), momentum flux (M) and volume flux (Q). Here, due to the limited size of the experiment, the integration between 0 and ∞ for some parameters is approximated by integrating up to the radial location at which the radial velocity v is less than 1% of the streamwise centreline velocity, i.e. $\overline{v} < 0.01u_{cl}$.

The last part of the paper will focus on dissipation and self-similar scalings. Hot-wire anemometry and 2D planar PIV are performed to study the flows both spatially and temporarily.

2 Methodology

The study is performed in a jet facility at Imperial College London, where the jet exits a nozzle with a sharp, top-hat mean velocity profile. Three different orifices (round, square and fractal, with fractal dimension $D_f = 1.5$ and 3 iterations) with identical exit open area are attached to the nozzle exit, as shown in Fig. 1. The fractal geometry was chosen to be the same as for the wake generator in Nedić et al. [8], which was shown to reduce the energy content of the large scale coherent structures. PIV was performed across the jet centreline between 0 and 23 equivalent diameters D_e at $Re_G = 3{,}000$ (exit velocity $U_e = 2.98\,\text{m/s}$) and $Re_G = 10{,}000$ ($U_e = 9.93\,\text{m/s}$). In total, 1,500 images were acquired per jet configuration. The field of view was $9\,D_e \times 6.7\,D_e$. An interrogation window of 16×16 pixels was used to process the

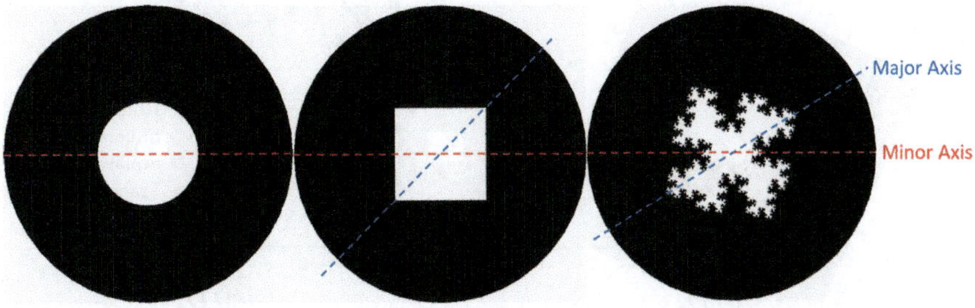

(a) Round (b) Square (c) Fractal

Fig. 1 Jet exits

images, with a 50% overlap, leading to a vector spacing of 0.03 D_e. The noncircular exits were studied both across the minor and the major axes.

$$D_e = \sqrt{Exit\ Area} \qquad U_e = \frac{Volumetric\ flow\ rate}{Exit\ Area} \qquad Re_G = \frac{U_e D_e}{\nu} \tag{4}$$

Following the preliminary results, highly resolved planar PIV (12 × 12 interrogation window, 50% overlap and spatial resolution of 3–4 Kolmogorov lenghtscale η) to investigate the velocity gradients was performed at various streamwise locations x in the near field (x/D_e from 2 to 5), in the intermediate field (x/D_e from 12 to 15, where the mean velocity starts becoming self-similar for all jets) and in the far field (x/D_e from 23 to 26, where the Reynolds stresses become self-similar). The testing campaign was concluded with single hot-wire anemometry in the near, intermediate and far-field at a 100 kHz of sampling rate, which was subsequently low-pass filtered at 30 kHz. A last set was measured in the far-field at $Re_G = 31{,}000$.

3 Results and Discussions

The two point correlation between the radial velocity component v in the streamwise direction x was analysed to investigate how the exit geometry affected the K–H vortex rings, as shown in Fig. 2a. It is shown that the fractal geometry suppresses the negative decorrelation, which is usually associated with the vorticity of the large scales, suggesting they have been broken-up. This is found up to 5 D_e, where the

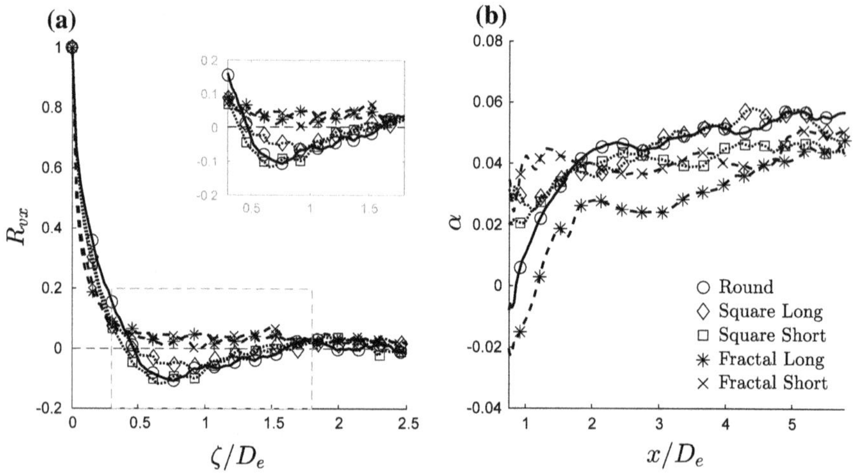

Fig. 2 Two point correlation at the shear layer (defined as the peak of $\langle u'^2\rangle^{1/2}$) at 3 D_e as a function of increasing streamwise distance ζ (**a**) and jet entrainment (**b**)

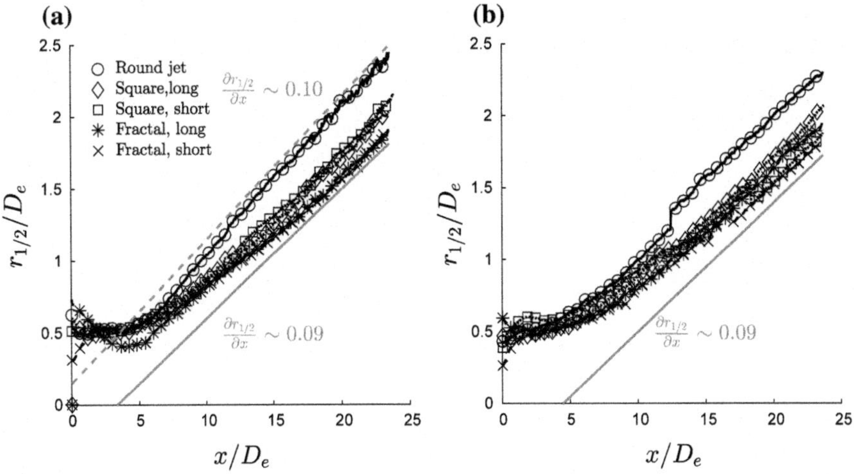

Fig. 3 Spreading rate $\partial r_{1/2}/\partial x$ of the various jets

fractal jet regains the decorrelation, even though it is significantly reduced in comparison to the other jets. The energy spectra calculated from the hot-wire data showed a reduced shedding energy and a less energetic spectrum for the fractal jet across the whole frequency range, compared to the round jet, whilst the square jet was less energetic than the round jet at low frequencies only. The autocorrelation of the streamwise fluctuation signal also showed a lack of periodicity in the oscillations, further confirming the break-up of coherent structures.

The different geometries also significantly affected entrainment and mixing. As shown in Fig. 2b, the calculated entrainment coefficient α is significantly reduced for the fractal case, while it is the highest for the round jet. In terms of mixing, at $Re_G = 3{,}000$, the round jet also has the highest spreading rate $\partial r_{1/2}/\partial x$, suggestive of a more rapid mixing, while the noncircular geometries have comparable values, as shown in Fig. 3a. At $Re_G = 10{,}000$, all jets have a comparable spreading rate (Fig. 3b), however due to the different location of the virtual origin resulting from the different initial conditions, the round jet is always the widest at the same streamwise location. It is therefore shown that the break-up of the coherent structures caused by the fractal geometry reduces the overall near-field entrainment and mixing, even though at a high enough Reynolds number the spreading rate is unaffected by the initial conditions. This point was also hypothesised by Townsend [10]. Moreover, in the near-field, the fractal jet shows a significantly reduced Reynolds shear stress.

The study is concluded with an analysis on the self-similar scaling exponents of the jets, to verify how the different initial conditions have affected them. Both the mean centreline velocity and jet half-width are expected to vary with a specific power law of x. Experimentally, it is found that all of the jets have the same macroscopic scaling exponents concerning centreline velocity and jet half-width. From the highly resolved PIV data, it is found for all jets that in the far field both mean velocity and

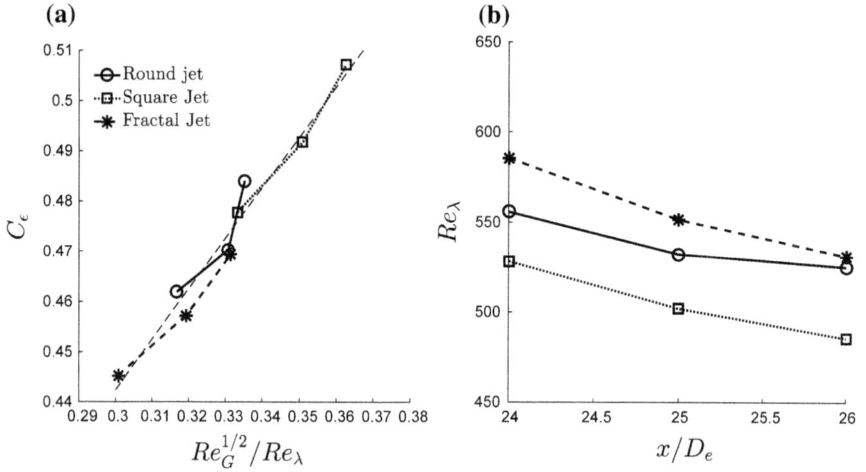

Fig. 4 C_ε and Re_λ at $Re_G = 31{,}000$ in the self-similar regime

Reynolds stresses are self-similar, however the local Reynolds number Re_λ has a decreasing trend, rather than being constant as it is expected from the classical theory, while the dissipation decreases more slowly than the expected $\varepsilon \sim x^{-4}$. Moreover, it is estimated that $Re_\lambda > 100$ for all the jet cases.

As explained by Goto and Vassilicos [4], this is the minimum Re_λ after which non-equilibrium dissipation appears. At $Re_G = 31{,}000$, Re_λ is found to be greater than 500. At this Reynolds number, C_ε is a function of $Re_G^{1/2}/Re_\lambda$ and Re_λ is decreasing as a function of x (Fig. 4). This indicates the presence of non-equilibrium dissipation for all the jet types between 24 and 26 D_e, where both mean velocity and Reynolds stresses are self-similar. Hence, it is shown that dissipation and the microscales of the flow take much longer than mean velocity and Reynolds stresses to regain the self-similar scalings and this is the case for both circular and noncircular jets.

4 Conclusions

This paper has explored the effect of initial conditions on the near-field flow physics and into the evolution of an axisymmetric turbulent jet, through the manipulation the near-field coherent structures with noncircular geometries. This has been achieved comparing different orifices with the same open area. It was shown that the large scale structures are suppressed by the fractal exit, while they are maintained for the square jet. Entrainment is shown to be reduced by the noncircular geometries and the round jet is proven to have the highest mixing and spreading rate for the same exit area. Moreover, in the near-field, the fractal geometry is shown to reduce fluctuations and turbulent kinetic energy significantly compared to square and round jet. However, in the intermediate and far-fields the behaviour of the jets is found comparable. Moreover, at $Re_G = 10{,}000$, the jets shows comparable spreading rates, even though the break-up of the coherent structures moves the jet virtual origin to a streamwise location further upstream. Non-equilibrium dissipation was also found in the self-similar region of the three jets, suggesting on one side that the evolution of an axisymmetric jet is independent of the initial conditions, but on the other that the microscales of the flow present signs of the initial conditions while mean velocity and Reynolds stresses are self-similar. Since usually the distance from the jet exit to reach self-similarity is evaluated based on the Reynolds stresses, it is shown this criteria is insufficient to reach the complete self-similarity of the scales of an axisymmetric turbulent jet.

References

1. R.A. Antonia, B. Pearson, Effect of initial conditions on the mean energy dissipation rate and the scaling exponent. Phys. Rev. E **62**(6), 8086–8090 (2000)
2. R.A. Antonia, Q. Zhao, Effect of initial conditions on a circular jet. Exp. Fluids **31**(3), 319–323 (2001)
3. W.K. George, The self-preservation of turbulent flows and its relation to initial conditions and coherent structures. Technical report, University at Buffalo, Buffalo, New York, USA (1989)
4. S. Goto, J.C. Vassilicos, Energy dissipation and flux laws for unsteady turbulence. Phys. Lett. Sect. A: Gen. At. Solid State Phys. **379**(16–17), 1144–1148 (2015)
5. E. Gutmark, F.F. Grinstein, Flow control with noncircular jets. Annu. Rev. Fluid Mech. **31**, 239–272 (1999)
6. R.J. Hearst, P. Lavoie, Decay of turbulence generated by a square-fractal-element grid. J. Fluid Mech. **741**, 567–584 (2014)
7. J. Nedić, O. Supponen, B. Ganapathisubramani, J.C. Vassilicos, Geometrical influence on vortex shedding in turbulent axisymmetric wakes. Phys. Fluids **27**, 035103 (2015)
8. J. Nedić, J.C. Vassilicos, B. Ganapathisubramani, Axisymmetric turbulent wakes with new nonequilibrium similarity scalings. Phys. Rev. Lett. **111**, 144503 (2013)
9. T. Shakouchi, S. Iriyama, Flow characteristics of submerged free jet flow from petal-shaped nozzle, in *4th International Conference on Jets, Wakes and Separated Flows* (2013)
10. A.A.R. Townsend, *The Structure of Turbulent Shear Flow*, 2nd edn. (Cambridge University Press, New York, 1976)
11. M. van Reeuwijk, J. Craske, Energy-consistent entrainment relations for jets and plumes. J. Fluid Mech. **782**, 333–355 (2015)
12. J.C. Vassilicos, Dissipation in turbulent flows. Annu. Rev. Fluid Mech. **47**, 95–114 (2015)

Study of Energetics in Drag-Reduced Turbulent Channel Flows

Davide Gatti, Maurizio Quadrio, Andrea Cimarelli, Yosuke Hasegawa and Bettina Frohnapfel

Abstract Changes in integral power budgets and scale energy fluxes as induced by certain active flow control strategies for turbulent skin-friction drag reduction are studied by performing Direct Numerical Simulation of turbulent channels. The innovative feature of the present study is that the flow is driven at Constant total Power Input (CtPI), which is a necessary enabling choice in order to meaningfully compare a reference unmanipulated flow with a modified one from the energetic standpoint. Spanwise wall oscillation and opposition control are adopted as model strategies, because of their very different control input power requirements. The global power budget show that the increase of dissipation of mean kinetic energy is not always related to drag reduction, while the preliminary analysis of the scale energy fluxes through the generalized Kolmogorov equation shows that the space- and scale properties of the scale energy source and fluxes are significantly modified in the near-wall region, while remain unaltered elsewhere.

1 Introduction

An important choice needs to be taken when setting up direct numerical simulations (DNS) of turbulent channel flows, regarding how the flow is driven through the channel. Two classic possibilities are to drive the flow at constant flow rate (CFR) or at

D. Gatti (✉) · B. Frohnapfel
Karlsruhe Institute of Technology, 76131 Karlsruhe, Germany
e-mail: davide.gatti@kit.edu

M. Quadrio
Politecnico di Milano, 20156 Milano, Italy
e-mail: maurizio.quadrio@polimi.it

A. Cimarelli
Università Degli Studi di Modena e Reggio Emilia, 41121 Modena, Italy
e-mail: andrea.cimarelli@unimore.it

Y. Hasegawa
University of Tokyo, Tokyo 153-8505, Japan
e-mail: ysk@iis.u-tokyo.ac.jp

R. Örlü et al. (eds.), *Progress in Turbulence VII*, Springer Proceedings in Physics 196, DOI 10.1007/978-3-319-57934-4_31

constant pressure gradient (CPG). While the different choices yield similar turbulent statistics for canonical flows [7], the main difference being, for instance, in the tails of the probability density function of wall shear stress, they have crucial implications on statistics of drag-reduced turbulent flows. For instance, with CFR drag reduction manifests as a reduction of friction but as an increase of bulk velocity with CPG. In neither case, the power transferred to the flow remains constant upon application of drag-reducing control nor so does the rate of production and dissipation of turbulent kinetic energy. Since the uncontrolled and drag-reduced flows differ energetically, it is difficult, if not impossible, to address the physics of drag reduction techniques from the energetic standpoint.

In this work, we exploit the recently-proposed constant total power input (CtPI) approach [4], in which the power transferred to the flow through pumping and imposition of a control is kept constant, to address how drag-reduction obtained via several wall-based strategies modifies energetic properties of turbulent channel flows. First, the effect of the control on the integral production and dissipation of mean and turbulent kinetic energy are computed [9]. Then, starting from the generalized form of the Kolmogorov equation [2, 6], the scale energy fluxes simultaneously occurring in the space of scales and in the physical space of wall-turbulent flows are preliminary studied to highlight differences among controlled drag-reduced and unmodified flows.

The structure of the paper is as follows. Section 2 describes the numerical method and procedures, as well as the control strategies adopted in the present study. In Sect. 3 the main results are presented and discussed, while Sect. 4 contains concluding remarks.

2 Numerical Method

Direct numerical simulation (DNS) of turbulent channel flows driven at CPI have been performed at a power-based Reynolds number, kept constant across all cases, of $Re_\Pi = U_\Pi h/\nu = 6500$, corresponding in the reference unmanipulated channel to $Re_\tau = u_\tau h/\nu = 199.7$ and $Re_b = U_b h/\nu = 3176.8$. In the previous definitions, U_Π is the bulk velocity of a laminar driven at the given power, u_τ and U_b are respectively the friction and the bulk velocity, h the channel semi-height and ν is the kinematic viscosity. Two active flow control strategies for turbulent skin-friction drag reduction, which require a control power input Π_c in order to be applied, have been considered, namely the spanwise-oscillating wall ([8]) and the oppoition control [1]. In such cases, the calculations are performed while keeping a Constant total Power Input (CtPI) [3] in time. The total power input Π_t is defined as the sum of the control power input Π_c and the pumping power Π_c, so that active control requires a fraction $\gamma = \Pi_c/\Pi_c$ of the total power to be spent for applying the control instead of directly pumping the flow.

The two control strategies of the present study (Fig. 1) have been selected due to their very different input power requirements, yielding different values of γ. The ocillating-wall forcing requires a significant amount of energy to operate, while

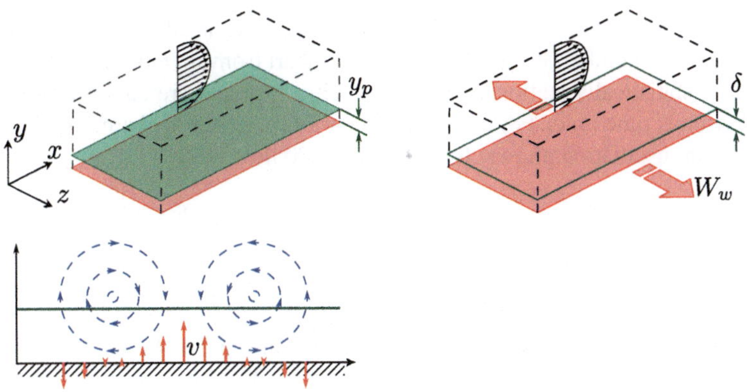

Fig. 1 Sketch of the two control strategies addressed in the present work. *Left* opposition control [1]. The wall-normal velocity is sensed at wall-parallel plane located a distance y_p from the wall and fed back at the wall as blowing and suction with opposite sign, so as to damp near-wall quasi streamwise vortices. *Right* spanwise wall oscillations [8]

the opposition control, which enforces a distributed vertical velocity v at the wall opposing the same component at a plane located at a prescribed wall distance y_p, requires minimal control power. The control parameters have been set in order to maximize the control performance, which in the CtPI framework means to maximize the increase in bulk mean velocity U_b (thereby decreasing the wall shear stress τ_w) at a fixed total power Π_t. This choice corresponds to an oscillating period of $T^+ = 125.5$ and a maximum spanwise wall velocity of about $W_w^+ = 4.5$ for the oscillating wall, and to a detection plane located at $y_p^+ = 13$ for the opposition control. Hereinafter, the superscript + denotes quantities that have been nondimensionalized by the actual friction velocity and kinematic viscosity. In these configurations, the oscillating wall achieved $Re_\tau = 186.9$ and $Re_b = 3268$ with a $\gamma = 0.098$, while the opposition control achieved $Re_\tau = 190.5$ and $Re_b = 3474$ with the much smaller $\gamma = 0.0035$.

The employed DNS solver is the one developed by Luchini and Quadrio [5], which uses a mixed spatial dicretization with Fourier series expansion in the two homogeneous span- and streamwise direction and fourth-order explicit compact finite differences in the wall-normal direction. The computational domain has a streamwise length of $L_x = 4\pi h$ and a spanwise width of $L_z = 2\pi h$. 256 Fourier modes are used to expand the velocity in the streamwise and spanwise direction before dealiasing (additional modes are added for avoiding aliasing), while the velocity is discretized in the wall-normal direction with 256 unevenly spaced points, in order to improve the resolution in the near-wall region. The corresponding spatial resolutions before dealiasing in viscous wall units are $\Delta x^+ = 9.8$, $\Delta z^+ = 4.9$, $\Delta y^+_{\min} = 0.47$ at the wall and $\Delta y^+ = 2.59$ at the channel centerline.

The governing equation are advanced in time with implicit temporal discretization for the viscous term and a third-order low-storage Runge-Kutta explicit scheme for the nonlinear terms. The time step is chosen to yield an averaged value of the Courant-Friedrichs-Levy number of 1.1. The calculations start from an initial condition where the flow is statistically stationary for the specific case and are advanced for about 25,000 viscous time units. For the oscillating-wall case, 200 field are saved for each of 8 different oscillation phases, for a total of 1600 flow fields.

3 Results

Table 1 summarizes the volume integrals of the inbound and outbound energy fluxes, normalized by the total power input Π_t. The flow is fed a pumping power Π_p and possibly a control input power Π_c, whose sum is by constraint constant. The flow dissipates the total power input directly as dissipation of the mean kinetic energy or via production of turbulent fluctuations as dissipation of turbulent kinetic energy. At the present low value of the Reynolds number, 59% of the total input power is dissipated directly by the mean flow, while the remaining 41% is "wasted" to produce turbulent kinetic energy and then dissipated as turbulent dissipation. When control is applied within the CtPi framework, all the global energy fluxes change in a way which strongly depends on the type of control considered or, in particular, on the control-to-total power ratio γ. In the case of the oscillating wall, for instance, 10% of the total power is used for applying the control, while only the remaining 90% is available for pumping. Nonetheless, the control results into an increase of the mean bulk velocity compared to the uncontrolled flow, which highlights how well the spanwise forcing class of control strategy performs, despite the high power requirement. The dissipation of mean kinetic energy decreases with the oscillating wall, in spite of the fact that U_b is increased, while the turbulent dissipation, which also accounts for the 10% of control input power considered as purely temporal velocity fluctuations, increases. The very opposite trend is observed for the opposition control, for which the dissipation of the mean kinetic energy is increased to 64% while the turbulent dissipation decreased to 36%.

Closer insights into the physics of such control techniques for turbulent drag reduction can be obtained by analysing the Kolmogorov equation generalized for anisotropic flows with mean shear. A detailed discussion of such equation is out of the scope of the present manuscript and the interested reader is demanded to the

Table 1 Integral power budget for the uncontrolled and controlled channels. MKE and TKE are abbreviations for Mean and Turbulent Kinetic Energy respectively. All values are given in fraction of the total power Π_t

	Reference	Oscillating wall	Opposition control
Pumping power Π_p	1.0	0.90	0.995
Control input power Π_c	0.0	0.10	0.005
MKE dissipation ϕ	0.59	0.54	0.643
TKE dissipation ε	0.41	0.46	0.357

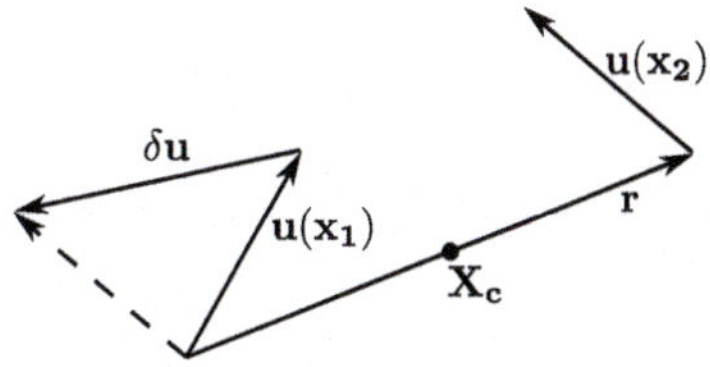

Fig. 2 Definition of the velocity difference, required to compute the second-order structure function. See text

discussion by Cimarelli, De Angelis and Casciola [6]. In the following we will discuss only the budget of the second order structure function $\langle \delta u^2 \rangle = \delta u_i \delta u_i$, where δu_i is the velocity difference (Fig. 2) between two points that are separated in space by a vector $\mathbf{r}$ and whose midpoint is located at the point $\mathbf{X_c}$. The angular brackets denote space and time averaging. In a channel flow, the second order structure function depends only on the three components r_i of the vector $\mathbf{r}$ and on the wall-normal coordinate Y_c of the midpoint $\mathbf{X_c}$. The second order structure function can be interpreted, according to its definition, as the amount of fluctuation energy at a scale r_i and at the spatial position Y_c and therefore will be called hereinafter scale energy. In the following only the properties only spanwise separations r_z and the wall-normal position Y_c will be addressed.

Figure 3 shows the maps of scale energy source term, positive when scale energy is produced and negative when dissipated, as colour maps. The vector field shows the fluxes of the scale energy in the $Y_c - r_z$ plane while the solid lines represent some field lines which originate at the singular point of the vector field. Only the near wall

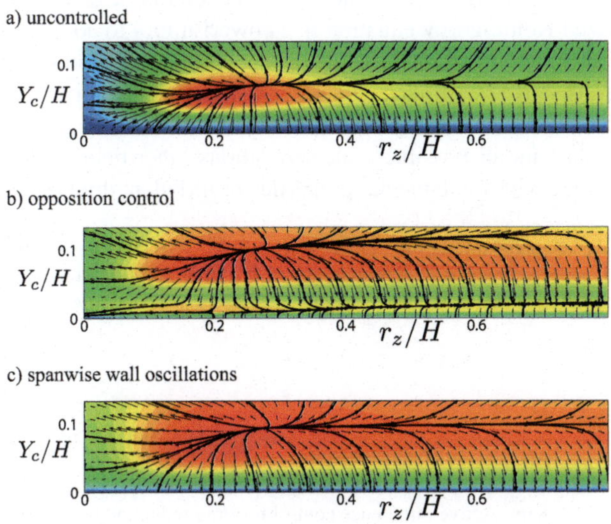

Fig. 3 Scale energy source map for an uncontrolled channel flow (**a**), for a channel modified by opposition control (**b**) and by spanwise wall oscillations (**c**). In all cases the total power is kept constant

region for the uncontrolled channel and for the channel manipulated via opposition control and spanwise wall oscillations are represented. Clearly, the morphology of scale energy production and dissipation is structurally modified by the control, also in this non trivial case in which the total power input to the system is kept constant. In the oscillating wall case, the peak scale-energy production decreases strongly and is located further from the wall compared to the reference case. The vertical shift of the peak scale energy production is observed also for the opposition control case. The presence of a wall vertical velocity causes the appearance of an intermediate region of positive scale energy production, located between the wall and the peak production. This intermediate region strongly modifies the morphology of the scale energy fluxes in the near-wall region.

4 Conclusions

The constant total power input (CtPI) approach is found to be a necessary and enabling step to address control-induced modifications of the energy transfer rates in turbulent flows without incurring in biases related to particular choice of scaling and normalizations of the results. The present study shows that no trivial general pattern can be found between the change of the global energy transfer rate and successful drag reduction, which in the CtPI results in an increase of bulk velocity compared to an unmanipulated flow. If the control input power is not accounted for in the turbulent dissipation, successful control results into a reduction of turbulent kinetic energy dissipation but not necessarily into an increase of mean kinetic energy dissipation.

The generalized Kolmogorov equation is a powerful tool to address the physics of turbulent drag reduction strategies from the energetic standpoint. The present preliminary results show a significant control-induced modification of both the scale energy production and scale energy fluxes in the near-wall region. Further analysis is required, considering all possible scale separations other than spanwise, to link the present evidence with fundamental properties of turbulent drag reduction.

Acknowledgements Support through the DFG project FR2823/5-1 is gratefully acknowledged. Computing time has been provided by the comutaional resource ForHLR Phase I funded by the Ministry of Science, Research and the Arts, Baden-Wrttemberg and DFG (Deutsche Forschungsgemeinschaft).

References

1. H. Choi, P. Moin, J. Kim, Active turbulence control for drag reduction in wall-bounded flows. J. Fluid Mech. **262**, 75–110 (1994)
2. A. Cimarelli, E. De Angelis, C.M. Casciola, Paths of energy in turbulent channel flows. J. Fluid Mech. **715**, 436–451 (2013)

3. B. Frohnapfel, Y. Hasegawa, M. Quadrio, Money versus time: evaluation of flow control in terms of energy consumption and convenience. J. Fluid Mech. **700**, 406–418 (2012)
4. Y. Hasegawa, M. Quadrio, B. Frohnapfel, Numerical simulation of turbulent duct flows at constant power input. J. Fluid Mech. **750**, 191–209 (2014)
5. P. Luchini, M. Quadrio, A low-cost parallel implementation of direct numerical simulation of wall turbulence. J. Comp. Phys. **211**(2), 551–571 (2006)
6. N. Marati, C.M. Casciola, R. Piva, Energy cascade and spatial fluxes in wall turbulence. J. Fluid Mech. **521**, 191–215 (2004)
7. M. Quadrio, B. Frohnapfel, Y. Hasegawa, Does the choice of the forcing term affect flow statistics in DNS of turbulent channel flow? Eur. J. Mech. B Fluids **55**, 286–293 (2016)
8. M. Quadrio, P. Ricco, Critical assessment of turbulent drag reduction through spanwise wall oscillation. J. Fluid Mech. **521**, 251–271 (2004)
9. P. Ricco, C. Ottonelli, Y. Hasegawa, M. Quadrio, Changes in turbulent dissipation in a channel flow with oscillating walls. J. Fluid Mech. **700**, 77–104 (2012)

Evolution of Vortex Formation in the Wake of Thin Flat Plates with Different Aspect-Ratios

Arman Hemmati, David H. Wood and Robert J. Martinuzzi

Abstract The effect of aspect-ratio (*AR*) on the formation and interaction of vortical structures in the wake of normal thin flat plates is examined at Re $=$ 1200 using an infinite span (2D) plate and rectangular plates ($AR = 1.0$, 1.6 and 3.2). The vortex shedding frequency significantly increased for $AR = 3.2$ compared to the 2D plate, while it dropped for $AR = 1.6$ and 1.0. The lowest frequency of vortex shedding was observed for $AR = 1.0$. Shear layers rolled to form vortices closer to the plate leeward surface at higher *AR*. The mean recirculation length was longer for the 2D plate compared to rectangular and square plates. The interaction of shear layers originating at the edges of the rectangular plates led to the formation of vortex loops in the wake. The wake appeared less organized for the lower *AR* plate (1.6) with a higher turbulence energy compared to $AR = 3.2$. The magnitude of turbulence kinetic energy was lowest at $AR = 3.2$ and it increased with decreasing *AR*. The mean drag coefficient was $\approx$2 for the 2D plate and $\approx$1 for $AR = 1$, which suggested major differences in the wake structures.

A. Hemmati (✉)
Department of Mechanical and Aerospace Engineering, Princeton University, Princeton, NJ, USA
e-mail: ahemmati@Princeton.edu

D.H. Wood · R.J. Martinuzzi
Department of Mechanical and Manufacturing Engineering, University of Calgary, Calgary, AB, Canada
e-mail: dhwood@ucalgary.ca

R.J. Martinuzzi
e-mail: rmartinu@ucalgary.ca

R. Örlü et al. (eds.), *Progress in Turbulence VII*, Springer Proceedings in Physics 196, DOI 10.1007/978-3-319-57934-4_32

1 Introduction

The formation and interaction of vortices in the wake of thin flat plates have significant implications on the fluctuating forces acting on the plate [2]. Thus, understanding the strength of vortex structures, their shedding process and their interactions in the wake is of great interest for various industrial applications including wind engineering, solar panel installations and propulsion performance of sharp-edge propeller fins. The most recent direct numerical simulations (DNS) of the flow past an infinite span normal thin flat plate [2] at Reynolds number of $Re = 1200$ and 2400 identified three distinct vortex shedding regimes based on the size and variation of lift and drag, organization of vortices, and periodicity of the vortex shedding. These regimes were identified as M, L and H for periods of moderate, low and high drag, respectively. Earlier studies [4, 6] had determined the existence of these regimes at a lower (250) and higher (18000) *Re*, respectively. However, these studies provided contradicting descriptions of the regimes due to qualitative nature of their identification [2].

The effect of side-edge vortices on the organization of the wake and the vortex shedding process for a thin normal flat plate of $AR = 3.2$ was investigated using DNS at $Re = 1200$ [3]. It was determined that the interaction of shear layers on four edges of the plate led to pre-mature detachment of vortices from shorter edges of the plate due to a secondary induced flow caused by the shedding of main roller vortices originating from the longer edges. This is referred to as the vortex "peel-off", which resulted in the formation of vortex loops in the wake [3]. Compared to the 2D plate wake, high *AR* rectangular plates had a smaller turbulence kinetic energy, a shorter mean recirculation length, and a larger vortex shedding frequency. Despite studying the differences in the vortex shedding mechanism between 2D and 3D plates [3], there has been no studies looking at the implications of *AR* on the vortex formation process for 3D plates.

This study looks at the effect of aspect-ratio on global flow variables, formation and interaction of vortices as well as the organization of the wake by comparing the flow past two rectangular ($AR = 3.2$ and 1.6), a square ($AR = 1.0$) and an infinite span (2D) thin flat plate, all placed normal to a uniform flow at $Re = 1200$. $AR = 1.6$ is typical of a modern photovoltaic (PV) module, 3.2 of an array of PV modules, and 1 is the lowest *AR* that needs investigating. The reason is that in the absence of significant Reynolds number effects, the flow over a flat plate of a certain *AR* will be identical to that of a plate of $1/AR$. Description of the numerical simulations are discussed in Sect. 2, followed by presentation of the results in Sect. 3. The main concluding remarks are discussed in Sect. 4.

2 Problem Description

The three-dimensional Navier–Stokes and continuity equations were solved using DNS at $Re = hU_0/\nu = 1200$, where U_0 is the freestream velocity, h is the plate chord, and ν is air viscosity. The second order backward Euler and the central difference method were used for temporal and spatial discritization of the flow field, respectively. An iterative Strongly Implicit Procedure (SIP) was used following the temporal discritization to solve the Poisson equation.

The computational domain extended 25h in the streamwise (x) direction with the inlet placed at $x = -5h$, and 16h in the chordwise (y) and spanwise (z) directions. Three stationary flat plates with the thickness of the smallest spatial grid was placed normal to the incoming uniform flow at the origin ($x = y = z = 0$) for each of the three simulations. The inlet boundary was uniform streamwise velocity ($u = U_0$ and $v = w = 0$) with zero pressure gradient. The side boundaries were assigned the freestream condition, and the boundary condition on plate surfaces was no-slip wall. The outlet boundary was assigned the Neumann outflow condition. The momentum component residual of 10^{-6} was used as the convergence criteria per timestep. The temporal grid was uniform and set small enough so that the convective (CFL) and diffusive numbers were maintained below 0.6.

Three unsteady simulations were completed using two rectangular plates with aspect-ratios ($AR = h/w$) 3.2 and 1.6, as well as a square plate ($AR = 1$) representing different arrays of PV modules. One can find more details about the simulations and verification of the results in comparison to existing numerical and experimental studies at [1].

3 Results and Discussion

The global flow parameters: vortex shedding Strouhal number ($St = f_s h/U_0$), mean recirculation length ($\overline{L_w}$), and mean drag ($\overline{C_d}$), are compared for the four cases in Table 1. These results demonstrate that the rectangular plates and the square plate have a significantly lower mean drag compared to the 2D plate, while the drag remains unchanged with changing AR for the rectangular plates. The behavior of $\overline{C_d}$ with respect to AR for rectangular plates is consistent with recent experimental studies [5]. Moreover, the vortex shedding frequency decreases with decreasing AR. $St = 0.317$ for $AR = 3.2$ and it decreases to 0.146 for the square plate ($AR = 1$). Compared to the 2D plate, however, St drops significantly at higher AR (3.2), but it gradually drops to comparable values for the square plate. The variation of St and $\overline{L_w}$ with changing AR is an indicator of the effect of AR on the vortex shedding process.

Table 1 The comparison of global flow variables in the wake of plates with different *AR*

AR	St	$\overline{C_d}$	$\overline{L_w}$
2D [2]	0.158	2.13	2.7
3.2 [3]	0.317	1.16	1.6
1.6	0.186	1.15	2.6
1.0	0.146	1.14	3.35

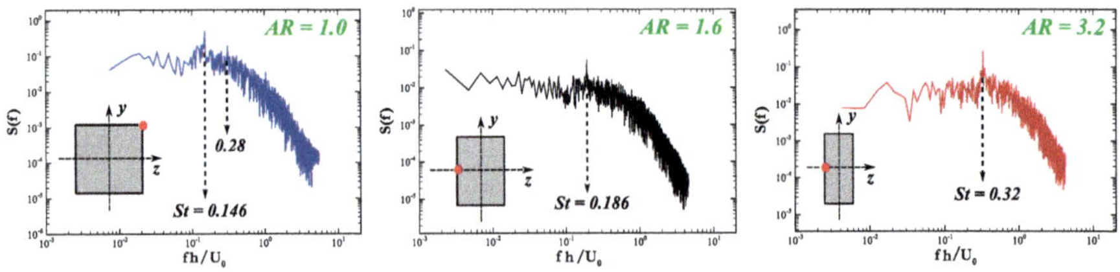

Fig. 1 The spectra of chordwise velocity fluctuations for different plate *AR*

The velocity spectra for $AR = 3.2$, 1.6 and 1 are shown in Fig. 1. Comparing spectral peaks for $AR = 3.2$ and 1.6, it is apparent that the vortex shedding has a lower frequency and is less energetic (in the frequency domain) at a lower *AR*. Moreover, there is only a single spectral peak identified at all locations in the wake of $AR = 3.2$, an example of which is presented in Fig. 1 associated with the dominant vortex shedding frequency. Although the same observations were valid for $AR = 1.6$, but the energy levels of the peak and *St* were both lower than that of $AR = 3.2$. Thus, there was a single shedding process for the rectangular plates. For $AR = 1$, however, there were energetic spectral peaks at either edges of the plate with comparable energy levels. There was also a second peak equaling the sum of the peaks determined at the two edges of the plate. Since this phenomenon was only observed at the plate corners, it could imply that there are two separate vortex shedding processes for the square plate.

The profiles of streamwise velocity along the wake centerline in Fig. 2 show smaller $\overline{L_w}$ for the higher *AR* plates, the velocity deficit inside the base vortex reduces with the *AR*. Moreover, the mean streamwise velocity asymptotically approaches that of the freestream for all three cases. The vorticity generation rate is observed to be similar for the three *AR*. Consequently, the strength of the shed vortices is expected to be inversely proportional to *St*. This observation is consistent with the increased entrainment at lower and higher values of the turbulence kinetic energy (k) observed for lower *AR*. The magnitude of k along the wake centerline (Fig. 2) is an indicator of

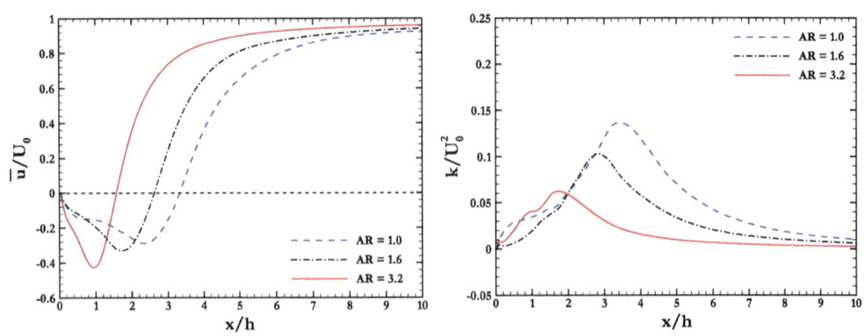

Fig. 2 The mean streamwise velocity and turbulence kinetic energy along the wake centerline

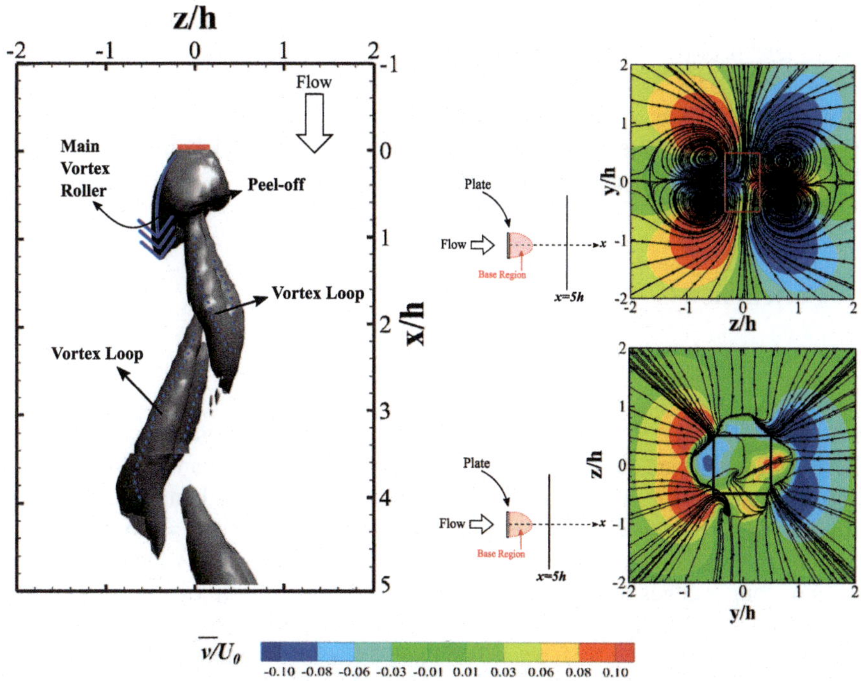

Fig. 3 The schematics of vortex shedding for $AR = 1.6$ (*left*), and contours of chordwise velocity for $AR = 1.6$ (*right- top*) and $AR = 1.0$ (*right-bottom*)

the changes on vortex formation process. First, the plateau of k at $0.8 < x < 1.2$ for $AR = 3.2$, which was attributed to a region of recurring vortex "peel-off" in previous studies [3], appears to shrink for the lower AR plate (1.6). Second, the magntidue of maximum k increases with reducing AR with the maximum k occurring for the square plate. Thus, the level of turbulence, which may be associated with more complex wake and vortex shedding, coincided with longer mean recirculation length and lower vortex shedding frequency. Third, the rate at which k rises to its maximum value for each AR is identical following the aforementioned plateaus for $AR = 3.2$ and 1.6, while the rate at which k drops following the end of the mean recirculation bubble is also similar for all AR. However, unlike the mean streamwise velocity, the asymptotic approach of all cases towards 0 is not at the same streamwise location.

The iso-surface plot of the λ_2 criterion for $AR = 1.6$ along with the contours of mean chordwise velocity and streamlines in Fig. 3 confirm that (1) there are 4 side-edge vortices formed in the wake, and (2) main vortex rollers are shed from longer edges while pre-mature structures are detached from shorter edges, as previously hypothesized for $AR = 3.2$ [3]. However, the contours of $AR = 1$ were different than those of the rectangular plates with side-edge-vortices missing and a change on the wake axis far downstream the plate at $x = 5h$. This suggested a very different vortex shedding process for the square plate compared to the rectangular plates, which were dominated by shedding of main vortex rollers from longer edges and side vortex "peel-off" from shorter edges [3].

This study is the first of its kind to evaluate the impact of *AR* on wake structures and vortex formation processes that had been previously investigated for $AR = 3.2$ [3] and 2D [2] plates positioned normal to the flow. It provided the first examination of the wake of a normal square flat plate.

4 Conclusions

The effect of aspect-ratio on formation and interaction of vortex structures in the wake of 3D normal flat plates was investigated at $Re = 1200$ using DNS. Comparing plates with AR =3.2, 1.6 and 1.0, the mean drag coefficient was monotonic with *AR*. There were significant differences in the shear layer extension prior to roll-up (vortex formation length), mean recirculation length and the vortex shedding frequency for different *AR*. Despite delays on the detachment of vortices for $AR = 1.6$, the vortex shedding mechanism remained the same for $AR = 1.6$ and 3.2, in which case shear layers on longer edges of the plate dominated the vortex roll-up followed by "peel-off" of vortices from the shorter edges. However, the vortex formation and shedding for the square plate ($AR = 1.0$) was completely different from that of the rectangular plates. The wake turbulence intensified as the *AR* decreased. The square plate had the largest maximum turbulence kinetic energy along the wake centerline.

Acknowledgements This study was completed at the University of Calgary, and it has received support from Natural Science and Engineering Research Council (NSERC) of Canada, the Alberta Innovations Technology Future (AITF), and the Schulich School of Engineering.

References

1. A. Hemmati, Evolution of large-scale structure in the wake of sharp-edge thin flat bodies. Ph.D. Thesis, University of Calgary, Calgary, AB, Canada (2016)
2. A. Hemmati, D.H. Wood, R.J. Martinuzzi, Characteristics of distinct flow regimes in the wake of an infinitely span normal thin flat plate. Int. J. Heat Fluid Flow. **62, Part B**, 423–436 (2016)
3. A. Hemmati, D.H. Wood, R.J. Martinuzzi, Effect of side-edge vortices and secondary induced flow on the wake of normal thin flat plates. Int. J. Heat Fluid Flow **61, Part A**, 197–212 (2016)
4. F.M. Najjar, S. Balachandar, Low-frequency unsteadiness in the wake of a normal flat plate. J. Fluid Mech. **370**, 101–147 (1998)
5. X. Ortiz, D. Rival, D. Wood, Forces and moments on flat plates of small aspect ratio with application to PV wind loads and small wind turbine blades. Energies **8**, 24–38 (2015)
6. S.J. Wu, J.J. Miau, C.C. Hu, J.H. Chou, On low-frequency modulations and three-dimensionality in vortex shedding behind a normal plate. J. Fluid Mech. **526**, 117–146 (2005)

Structure of Turbulence in a Flow Around a Rectangular Cylinder

Adriano Leonforte, Andrea Cimarelli and Diego Angeli

Abstract The behaviour of the flow over a finite blunt plate with square leading and trailing edge corners at moderate Reynolds number is studied by means of a Direct Numerical Simulation. The chord-to-thickness ratio of the plate is 5 and the Reynolds number is $Re = U_\infty \cdot D/\nu = 3 \times 10^3$ where U_∞ and D are the free-stream velocity and the thickness, respectively. The flow separates at the leading edge corner developing in a strong free-shear. The flow reattaches on the solid surface upstream the trailing edge and evolves in typical large-scale shedding beyond it. To the authors knowledge, this is the first time that high-fidelity three-dimensional data are produced to analyze in detail the flow features of such a system. Preliminary results on the flow topology will be presented in this work. In particular, the streamlines of the mean flow and the instantaneous three-dimensional turbulent structures via λ_2 vortex criterion will be examined.

1 Introduction

Rectangular cylinders appear in several engineering configurations, ranging from civil engineering, such as long-span bridge decks or high-rise buildings, to automotive engineering. As its chord-to-thickness ratio is varied from zero to infinity, the rectangular cylinder encompasses the range of bluff bodies from a flat plate normal to the flow, to a square cylinders, and, finally, to a flat plate parallel to the flow [8]. A detailed study on the unsteady behaviour of separating and reattaching flow was presented by Cherry et al. [2]. Two main types of instabilities are involved: the Kelvin-Helmholtz instability present in the leading edge shear

A. Leonforte (✉) · A. Cimarelli · D. Angeli
DISMI, University of Modena and Reggio Emilia, Via Amendola 2, 42122 Reggio Emilia, Italy
e-mail: adriano.leonforte@unimore.it

A. Cimarelli
e-mail: andrea.cimarelli@unimore.it

D. Angeli
e-mail: diego.angeli@unimore.it

R. Örlü et al. (eds.), *Progress in Turbulence VII*, Springer Proceedings in Physics 196, DOI 10.1007/978-3-319-57934-4_33

layer and the large-scale shedding after the trailing edge. Furthermore, a series of interacting mechanisms from the impingement region to the free-shear layer are observed and found to be at the basis of the transition to turbulence. A detailed investigation into the nature of two-dimensional and three-dimensional turbulent structures can be found in Hourigan et al. [3] and Sasaki and Kiya [6]. However, a clear picture of the self-regeneration turbulent mechanisms is still missing and needs a further assessment. Concerning the wake, a *large-scale* vortex shedding is present and has been studied by many researchers. Among others, Nakamura et al. [4] and Ohya et al. [5] by using both experiments and simulations, observed an interesting feature, i.e. that the Strouhal number based on plate length was approximately equal to $0.6n$, where n is an integer which represents the number of vortices simultaneously present on the plate and, hence, is related with the length of the rectangular prism's chord. Despite the large interest on the flow configuration and the lack of clarity on the physics behind the transition to turbulence, at present no Direct Numerical Simulations (DNS) at sufficiently high Reynolds number has been performed. In contrast with experiments, a DNS approach would allow us to have access at the multi-dimensional features determining the transition to turbulence in such a flow. Actually, an attempt to investigate the nature of this complex flow via DNS was proposed in the past. Tamura et al. [7] approached the problem by means of a finite difference technique at high Reynolds number equal to 10^4. However, the grid resolution adopted was not fine enough to capture the smallest scales of motion. Hence, the simulation reported appears to be more an implicit Large Eddy Simulation than a DNS. More recently Hourigan et al. [3] proposed a more accurate analysis through the spectral-element method. However, the DNS data reported are concerning very low Reynolds numbers, namely $Re = 350$ and $Re = 400$, where a fully developed turbulent state at this Reynolds numbers is not achieved.

The lack of an accurate DNS database at sufficiently large Reynolds numbers is at the basis of the present work. Indeed, our aim is to present for the first time high-fidelity data for such a flow at moderate Reynolds number. A preliminary analysis of these data in terms of mean flow feature and turbulent structures will be presented. This study represents the first attempt towards a complete understanding of the self-sustaining mechanism of turbulence in such a flow configuration.

2 Numerical Approach

We performed a well-resolved DNS for the flow around a plate with chord-to-thickness ratio 5. The considered Reynolds number is $Re = U_\infty D/\nu = 3 \times 10^3$ where U_∞ is the free-stream velocity and D is the thickness of the rectangular cylinder. The dimensions of the rectangular plate are $(L_x, L_y, L_z) = (5D, D, 5D)$ and the size of the domain is $(\mathscr{D}_x, \mathscr{D}_y, \mathscr{D}_z) = (112D, 50D, 5D)$. The continuity and momentum equations are discretized by a finite volume method [9] through central schemes in space and a backward Euler scheme in time both accurate at the second order. The PISO algorithm is employed for solving the pressure–velocity linked equations.

A flat velocity profile U_∞ without any superimposed disturbance is applied at the inlet, periodic boundary conditions are imposed in the spanwise direction while in the vertical directions and in the outlet freestream boundary conditions are applied. The mesh is made up of structured hexahedral blocks, about 1.5×10^7 volumes, with different spanwise resolution, changing in a stepwise manner from $Nz = 9$ (far from the body) to $Nz = 144$ (close to the body). The corresponding resolution is $(\overline{\Delta x^+}, \overline{\Delta y^+_{min}}, \overline{\Delta z^+}) = (6.1, 0.31, 5.41)$ where $\overline{(\cdot)}$ denotes the average in the streamwise direction and the superscript + stands for non-dimensionalization in viscous units. Statistics are computed after the initial transient taking advantage of the statistical stationarity of the flow field in time, by means of $N = 317$ samples, the statistical homogeneity in the spanwise direction and at last the symmetries in the vertical direction.

3 Results

In the following, we report a preliminary analysis of the DNS data set in terms of single-point statistics and identification of turbulent structures.

In Fig. 1 the behaviour of the mean flow $(U, V, 0)$ in the $(x - y)$-plane is shown with streamlines. The sharp leading edge corner fixes the position for the separation of the incoming undisturbed flow. The separated flow is found to reattach, on average, before the trailing edge. Hence, a large recirculating bubble takes place, green lines of Fig. 1, extending in the streamwise and vertical directions up to $x/D \approx 3.65$ and $y/D \approx 0.5$, respectively. At the reattachment point, two boundary layers take origin, one propagating upstream and the other towards the trailing edge. The last one finally detaches at the trailing edge and develops in the wake. The mean dimensions of the wake vortex are highlighted in Fig. 1 with cyan lines. The reverse boundary layer is firstly subjected to a favorable gradient pressure near the reattaching zone, see Fig. 2 where the wall-profiles of the pressure and friction coefficients are reported. Then, by further moving upstream, for $x/D < 1.8$, the strong favorable pressure gradient reverses, see again Fig. 2. As a consequence, the reverse boundary layer detaches, thus leading to a secondary recirculating bubble, rotating counterclockwise, extending for $0.4 < x/D < 1.4$ and $0 < y/D < 0.08$, see red lines in Fig. 1.

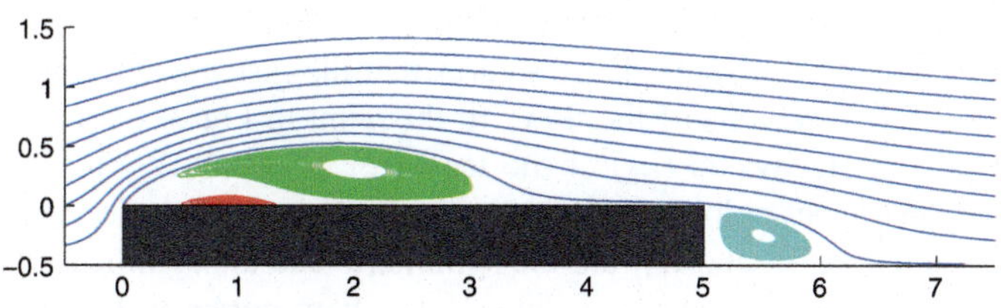

Fig. 1 Streamlines of the mean flow $(U, V, 0)$ in the $(x - y)$-plane

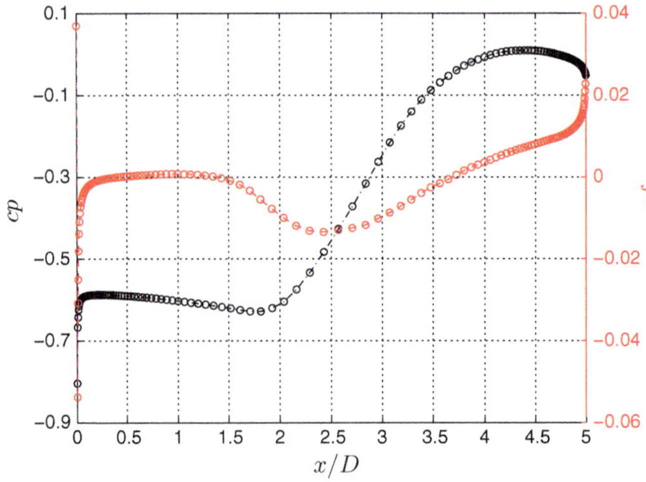

Fig. 2 Pressure coefficient at the wall (*black lines*) and skin friction coefficient (*red lines*) along the streamwise direction

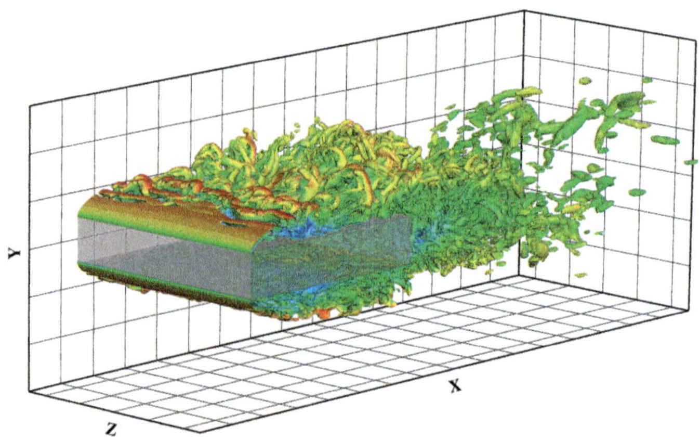

Fig. 3 Identification of turbulent structures via $\lambda_2 = -2$ *colored* by u (streamwise velocity)

The complex physical features characterizing the separated and reattaching flow can be highlighted by analysing the structures that the instantaneous velocity field takes in the different regions of the flow. To this aim, in Fig. 3, the regions where the second largest eigenvalue (λ_2) of the tensor $S_{ik}S_{kj} + \Omega_{ik}\Omega_{kj}$ is negative, are shown with iso-surfaces colored by the magnitude of the streamwise velocity for a value $\lambda_2 = -2$. Here, S_{ij} and Ω_{ij} are the symmetric and antisymmetric part of the velocity gradient tensor. As shown in Fig. 3, the sharp corner at the leading-edge fixes the location of the boundary layer detachment of the incoming flow. In the very first part of the shear layer, $x/D < 0.3$, the flow is laminar, then the spatially

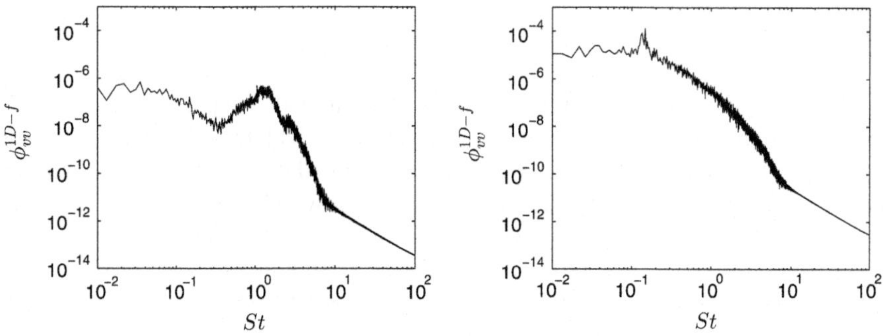

Fig. 4 Time-spectra of vertical velocity ϕ_{vv}^{1D-f} as a function of the Strouhal number $St = fD/U_\infty$ evaluated in the shear layer (*left*) and in the wake (*right*)

developing shear layer grows and, through instability and transitional phenomena breakdown to turbulence. The Kelvin-Helmholtz instability develops first, leading to the formation of two-dimensional spanwise vortices. Then transition to turbulence takes place for $x/D > 0.5$. Under the effect of the mean shear, still strong at this streamwise location, perturbations of the flow field lead to the lift up and stretching of the primary spanwise vortices, thus forming hairpin-like structures. These sparsely distributed structures, grow, evolve and decay into multi-scale turbulent fluctuations. The statistical footprint of this turbulent motion can be exploited by means of the time-spectrum of the vertical velocity ϕ_{vv}^{1D-f}. As shown in the left-hand plot of Fig. 4, the amplified turbulent structures in the shear layer are characterized by a well-defined time scale corresponding to a Strouhal number, $St = fD/U_\infty \approx 1.3$ where f is the frequency. Some of these fluctuations move back through the reverse boundary layer, see the blue isosurface in Fig. 3. These structures moving upstream toward the leading edge are firstly stretched in the streamwise direction due to the re-acceleration imposed by the pressure field, then after passing the mean primary vortex core, for $x/D < 1.8$, they start to decelerate, due to an adverse pressure gradient. This adverse pressure gradient leads also to a partial re-orientation of vortex tubes from the streamwise direction to the spanwise one. A deeper study of these structures is required, since, they take part of the feedback mechanism triggering the leading-edge shear layer. Concerning the boundary layer moving downstream towards the trailing edge, the relatively small turbulent structures are encompassed by oscillatory large scale motions reminiscent of the laminar von Kármán instability. Indeed, as shown in the right-hand plot of Fig. 4, the time-spectrum is filled up by the developed multi-scale structures of the trailing edge but, at the same time, a well-defined peak at small frequency relative to the large-scale shedding, $St = 0.14$, is also present.

4 Conclusions

In summary, given the lack of high-fidelity numerical data, we performed a DNS of the flow around a rectangular cylinder with chord-to-thickness ratio equal to 5 for $Re = 3 \times 10^3$. The statistical analysis of these data will be used as a benchmark for the scientific community studying this canonical flow [1] and will provide useful insights on the more general physics of self-sustaining turbulence in separated and reattaching flows. Here, we reported preliminary results in terms of mean velocity field and time-spectra. These statistical quantities clearly identify and characterize the main recirculating bubble and the secondary one in terms of length and time scales. On the other hand, the analysis of the turbulent structures reveals the presence of streamwise vortices and high- and low-speed streaks as a result of the presence of hairpin-like structures in the leading-edge free-shear layer.

References

1. L. Bruno, M.V. Salvetti, F. Ricciardelli, Benchmark on the aerodynamics of a rectangular 5: 1 cylinder: an overview after the first four years of activity. J. Wind Eng. Ind. Aerodyn. **126**, 87–106 (2014)
2. N.J. Cherry, R. Hillier, M.E.M.P. Latour, Unsteady measurements in a separated and reattaching flow. J. Fluid Mech. **144**, 13–46 (1984)
3. K. Hourigan, M.C. Thomson, B.T. Tan, Self-sustained oscillations in flows around long blunt plates. J. Fluid Struct. **15**, 387–398 (2001)
4. Y. Nakamura, Y. Ohya, H. Tsuruta, Experiments on vortex shedding from flat plates with square leading and trailing edges. J. Fluid Mech. **222**, 437–447 (1991)
5. Y. Ohya, Y. Nakamura, S. Ozono, H. Tsuruta, R. Nakayama, A numerical study of vortex shedding from flat plates with square leading and trailing edges. J. Fluid Mech. **236**, 445–460 (1992)
6. K. Sasaki, M. Kiya, Three-dimensional vortex structure in a leading-edge separation bubble at moderate Reynolds numbers. ASME J. Fluids Eng. **113**, 405–410 (1991)
7. T. Tamura, Y. Itoh, K. Kuwahara, Computational separated-reattaching flows around a rectangular cylinder. J. Wind Eng. Ind. Aerodyn. **50**, 9–18 (1993)
8. B.T. Tan, M.C. Thompson, K. Hourigan, Flow past rectangular cylinders: receptivity to trtansverse forcing. J. Fluid Mech. **515**, 33–62 (2004)
9. H.G. Weller, G. Tabor, H. Jasak, C. Fureby, A tensorial approach to computational continuum mechanics using object-oriented techniques. Comput. Phys. **12**, 620–631 (1998)

Application of Sponge Boundary Conditions to Large-Eddy Simulation of Multiple Thermal Plumes

Chandra Shekhar Pant and Amitabh Bhattacharya

Abstract While numerically simulating multiple thermal plumes, instabilities arise at the inflow/outflow/convective (IOC) boundaries due to the inability of IOC boundary conditions to effectively advect the flow structures out of the computational domain. To address this issue, Large Eddy Simulation of multiple thermal plumes is carried out with a novel viscous sponge layer, recently formulated by Pant and Bhattacharya (Comput Fluids 134:177–189, 2016, [5]). We validate our results with available literature, and we also present the effect of sponge layer on vortex rings (produced by thermal bubbles) traveling at an oblique angle to the outflow boundary. It is concluded from this study that the sponge layer does not appreciably affect the vortex rings outside the layer. The vortex rings do get smoothened out inside the sponge layer, which ensures a smooth convective outflow velocity for the eddies, and leads to stable LES of thermal plumes.

1 Introduction

Pure multiple thermal plumes commonly exist in both natural and industrial processes. Interaction of multiple jets/plumes can be seen as a simplified model for interacting atmospheric clouds. In the context of deep clouds, Baines and Keffer [1] experimentally studied multiple jets of air at ambient temperature. Kaye and Linden [4] experimentally studied interaction of two axi-symmetric turbulent plumes and proposed a model. Recently, Cenedese and Linden [2] studied the dynamics of the two coalescing plumes and proposed a net (or “effective”) entrainment constant for the multiple plumes. Numerical simulation of multiple plumes remain challenging because of the inadequacy of the proper boundary conditions. In this manuscript we performed Large Eddy Simulation (LES) of multiple plumes using the sponge boundary condition proposed by Pant and Bhattacharya [5]. To illustrate the effect of the viscous sponge layer on the flow structures, we also present results on isolated vortex rings impinging the sponge layer at an angle.

C.S. Pant · A. Bhattacharya (✉)
Indian Institute of Technology Bombay, Powai, Mumbai 400076, India
e-mail: bhattach@gmail.com

R. Örlü et al. (eds.), *Progress in Turbulence VII*, Springer Proceedings in Physics 196, DOI 10.1007/978-3-319-57934-4_34

2 Flow Domain, Governing Equations and Viscous Sponge Layer Formulation

A cuboidal computational domain is used, with dimension of $L_x \times L_y \times L_z$. Figure 1a shows the two dimensional schematic of the computational box. The complete non-dimensional equations for velocity field $\mathbf{u}$, pressure P and temperature, T used in this work are given as:

$$\nabla \cdot \mathbf{u} = 0 \tag{1}$$

$$\frac{\partial \mathbf{u}}{\partial t} + \mathbf{u} \cdot \nabla \mathbf{u} = -\nabla P + (\mathrm{Ri})T\mathbf{j} - \nabla \cdot \sigma^t + \mathbf{b}^s \tag{2}$$

$$\frac{\partial T}{\partial t} + \mathbf{u} \cdot \nabla T = - \nabla \cdot \mathbf{Q} \tag{3}$$

Here, $\mathrm{Ri} = \beta \Delta T$ refers to the Richardson number, which is used to couple temperature and velocity evolution equations. $\mathbf{j}$ is the unit vector in the positive y direction, σ^t is the stress tensor due to unresolved turbulent scales and $\mathbf{Q}$ is the heat flux vector. Terms σ^t and $\mathbf{Q}$ are modeled using standard Smagorinsky-Lilly model [6]. $\mathbf{b}^s$ is the body force due to artificial viscosity in the sponge, given by:

$$b_i^s = H(\mathbf{x})\nu^s \frac{1}{h_1 h_2 h_3}\left[\frac{\partial}{\partial \xi_1}\left(\frac{h_2 h_3}{h_1}\right)\frac{\partial v_i}{\partial \xi_1} + \frac{\partial}{\partial \xi_3}\left(\frac{h_1 h_2}{h_3}\right)\frac{\partial v_i}{\partial \xi_3}\right] \tag{4}$$

the term H($\mathbf{x}$) is 1 inside the sponge layer and 0 outside the domain, ν^s is the artificial viscosity. This novel sponge layer formulation allows a discontinuous jump in viscosity, so that the flow structures get dissipated inside the sponge layer. The detailed

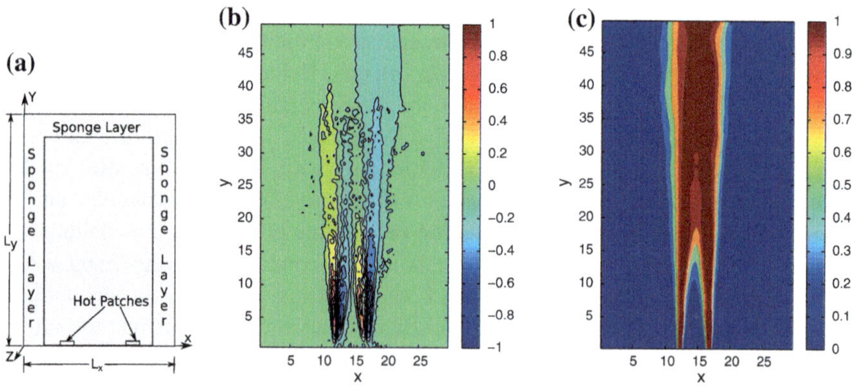

Fig. 1 **a** Schematic of computational domain. Time contour plots corresponding to case A1 of **b** average vorticity, $\langle \omega_3 \rangle$, and **c** average temperature field $\langle T \rangle(\mathbf{x})$ normalized with respect to temperature at $x = 2L_x/5$, $z = L_z/2$ ($x = 2L_x/5$, $z = L_z/2$ refers to the location of one of the hot patch at the bottom of computational domain)

formulation and solution procedure is explained in [5]. In this manuscript we assume that the flow has infinite Reynolds number, therefore Richardson number (Ri) is the only the free non-dimensional parameter.

3 Results

3.1 Simulation of Multiple Plumes

Two hot patches ($T = 1$) are introduced at the bottom of domain to simulate multiple thermal plumes, as shown in the Fig. 1a. No penetration boundary condition is used for the velocity field on the bottom plane ($y = 0$). The other boundary conditions are same as that explained in [5]. Two set of simulations (cases A1 and A2) are performed with different sponge layer viscosity ν_s. The parameters of the simulations are given in Table 1. In Fig. 2, the volume averaged kinetic energy is plotted against time for case A1 and A2, and it appears that both the simulations are stable for a long time. The contour plots of vorticity and normalized temperature field averaged in time are shown in Fig. 1b and c respectively. The intermixing between the two plumes and irrotationality of ambient velocity field is clearly evident from Fig. 1b. Figure 1c clearly shows that two thermal plumes develop independently until the axial distance of 10. Thereafter, the two thermal plumes start interacting with each other, and, beyond a certain height, become indistinguishable. These features of different flow regimes have been characterized by Cenedese and Linden [2].

Table 1 Simulation parameters for multiple thermal plume

Case	$L_x \times L_y \times L_z$	$N_x \times N_y \times N_z$	ν^s
A1	$30 \times 50 \times 30$	$113 \times 190 \times 113$	0.02
A2	$30 \times 50 \times 30$	$113 \times 190 \times 113$	0.05

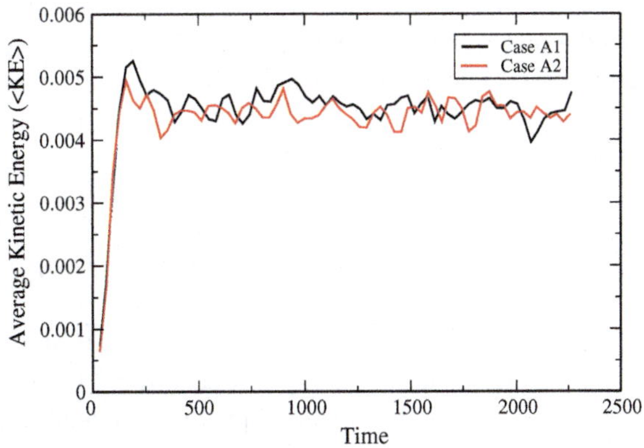

Fig. 2 Volume averaged kinetic energy $\frac{1}{2L_xL_yL_z}\int_{\mathscr{D}} u_iu_i d\mathbf{x}$ versus time for cases A1 and A2

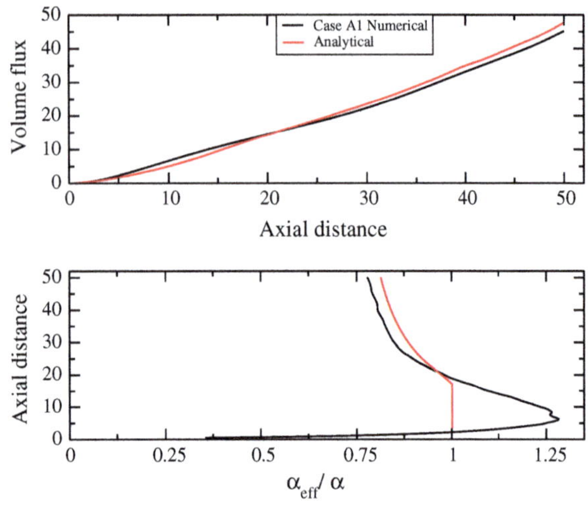

Fig. 3 Variation of volume flux and ratio of entrainment coefficient with axial distance

The total volume flux of two plumes (Q) and ratio of entrainment coefficients α_{eff}/α of the present LES simulations are validated against the analytical results of Cenedese and Linden [2] (shown in Fig. 3). Here F is the combined buoyancy flux for two plumes, α is the entrainment for single plume and α_{eff} is the net entrainment rate coefficient for the two interacting plumes. Unlike the analytical results, the ratio α_{eff}/α is significantly larger than 1 for $y < 20$ (y is the axial distance), which could be due to the relatively small value of y/D (D is the diameter of the patch). However, for $y > 20$, the agreement between simulation and analysis is reasonably good.

3.2 *Effect of the Sponge Layer on Flow Structures*

The LES of multiple thermal plumes is numerically stable, mainly due to the smoothening of flow structures near the convective outflow boundary. To illustrate the effect of the sponge layer on flow structures, we have carried out LES simulations of a thermal bubble, which then gives rise to a disintegrating vortex ring. The initial temperature field inside the bubble is spherically symmetric, with radius $r_c = 1.25$, centered at $(x_c, y_c, z_c) = (40, 1.25, 5)$. The initial temperature field is adapted from [3] and given by:

$$T = \begin{cases} 0, & r > r_c \\ A[1 + \cos(\pi \times r/r_c)], & r \leq r_c \end{cases}$$

where, $r = \sqrt{(x - x_c)^2 + (y - y_c)^2 + (z - z_c)^2}$ and $A = 0.5$.

Acceleration due to gravity has been tilted by 45 degrees to $-g(\mathbf{i} + \mathbf{j})/\sqrt{2}$ for these cases, so that the bubble impinges at an angle to the sponge layer. Velocity at

Table 2 Simulation parameters for vortex ring-sponge layer interactions

Case	$L_x \times L_y \times L_z$	$N_x \times N_y \times N_z$
B1	$80 \times 20 \times 10$	$400 \times 150 \times 100$
B2	$80 \times 40 \times 10$	$400 \times 300 \times 100$

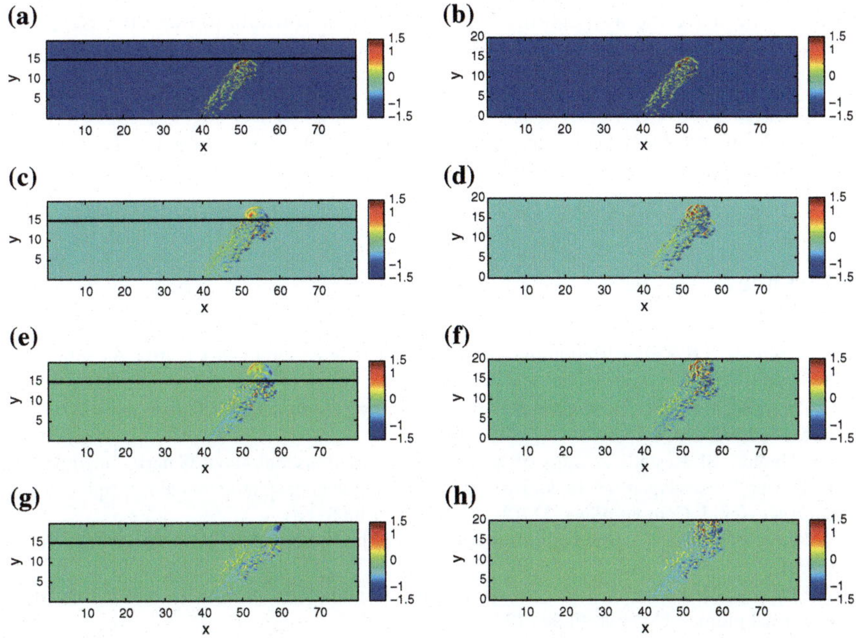

Fig. 4 Contour plot of vorticity ω_3 for case B1 at $z = L_z/2$: **a**, **c**, **e**, **g** and for case B2: **b**, **d**, **f**, **h**. Snapshots at the same row correspond to exactly the same time instant for both cases. A reference line has been drawn at $y = 15$, indicating the starting point of the top sponge layer in case B1

the bottom wall is set zero. Neumann boundary condition is used for top and lateral boundaries. The value of Ri is set 0.25. Table 2 lists the size of the computational box for the case B1 and case B2. The domain height L_y of case B1 is half that of case B2, and, correspondingly, the sponge layer height is 5, ranging from $y = 15$ to $y = 20$, which is again half compared to the case B2 (sponge height is 10, extends from $y = 20$ to $y = 30$). The domain in lateral and tangential directions is same for both cases. The grid resolution for both the cases is same. The iso-contours of vorticity at same time instants are compared for case B1 and case B2. These plots are shown before and after the penetration of vortex ring into the sponge layer $y = 15$ (Fig. 4). For case B1, outside the sponge layer (i.e. for $y < 15$) the flow structures look identically similar to case B2 (where, effectively, sponge layer is not present). At the same time, the sponge layer ensures that the flow structures are smooth near the convective outflow boundary, leading to stable simulations.

4 Conclusion

We have demonstrated that the novel viscous sponge layer scheme by [5] leads to stable LES for multiple plumes, allowing for long simulations over which statistics can be gathered. We found reasonable agreement of the net entrainment rate coefficient with analytical expressions by [2]. However, a taller computational domain may be needed to improve the agreement. The sponge layer formulation can be useful for LES of atmospheric flows, where multiple thermal plumes often interact with each other.

Acknowledgements AB acknowledges support from I.I.T. Bombay seed grant (Project Code 12IRCCSG020) for this work.

References

1. W.D. Baines, J.F. Keffer, Entrainment by a multiple source turbulent jet. Adv. Geophys. **18**, 289–298 (1975)
2. C. Cenedese, P. Linden, Entrainment in two coalescing axisymmetric turbulent plumes. J. Fluid Mech. **752**, R2 (2014)
3. F.X. Giraldo, M. Restelli, A study of spectral element and discontinuous Galerkin methods for the Navier–Stokes equations in nonhydrostatic mesoscale atmospheric modeling: equation sets and test cases. J. Comput. Phys. **227**(8), 3849–3877 (2008)
4. N. Kaye, P. Linden, Coalescing axisymmetric turbulent plumes. J. Fluid Mech. **502**, 41–63 (2004)
5. C.S. Pant, A. Bhattacharya, A viscous sponge layer formulation for robust large eddy simulation of thermal plumes. Comput. Fluids **134**, 177–189 (2016)
6. J. Smagorinsky, General circulation experiments with the primitive equations: I. The basic experiment. Mon. Weather Rev. **91**(3), 99–164 (1963)

Heat Transport in Horizontal and Inclined Convection

Olga Shishkina

Abstract We discuss three classical paradigmatic systems of thermally driven flows: Rayleigh–Bénard convection, where a fluid is confined between a heated bottom plate and a cooled top plate, horizontal convection, where the fluid is heated at one part of the bottom and cooled at some other part, and vertical convection, where the fluid is confined between two differently heated isothermal vertical plates. Rayleigh–Bénard and vertical convection can be also considered as limiting cases of so-called inclined convection. For these systems we study how the heat and momentum transport, which is represented by the Nusselt number and Reynolds number, scales with the main governing parameters of the system, which are the Rayleigh number and Prandtl number. We show that different boundary conditions generally lead to different scaling diagrams in the Prandtl–Rayleigh plane. For laminar vertical convection the scalings can be derived from the boundary layer equations, see Shishkina (Phys Rev E 93:051102, 2016, [8]). In the case of horizontal convection, the scalings can be derived from the analysis of the boundary-layer and bulk contributions of the kinetic and thermal dissipation rates, see Shishkina et al. (Geophys Res Lett 43:1219–1225, 2016, [5]). Here we summarize some previous results and discuss the applicability of the developed theory to global ocean circulation.

1 Introduction

Thermally driven flows are ubiquitous in nature. The classical paradigmatic systems to study such flows are Rayleigh–Bénard convection (RBC) [1, 2], where a fluid is confined between a heated bottom plate and a cooled top plate, horizontal convection [3–6], where the fluid is heated at one part of the bottom and cooled at some other part, and vertical convection, where the fluid is confined between two differently heated isothermal vertical plates [7–9] (see Fig. 1). Rayleigh–Bénard and vertical convection can be also considered as limiting cases of so-called inclined convection [13].

O. Shishkina (✉)
Max Planck Institute for Dynamics and Self-Organization,
Am Fassberg 17, 37077 Göttingen, Germany
e-mail: Olga.Shishkina@ds.mpg.de

R. Örlü et al. (eds.), *Progress in Turbulence VII*, Springer Proceedings
in Physics 196, DOI 10.1007/978-3-319-57934-4_35

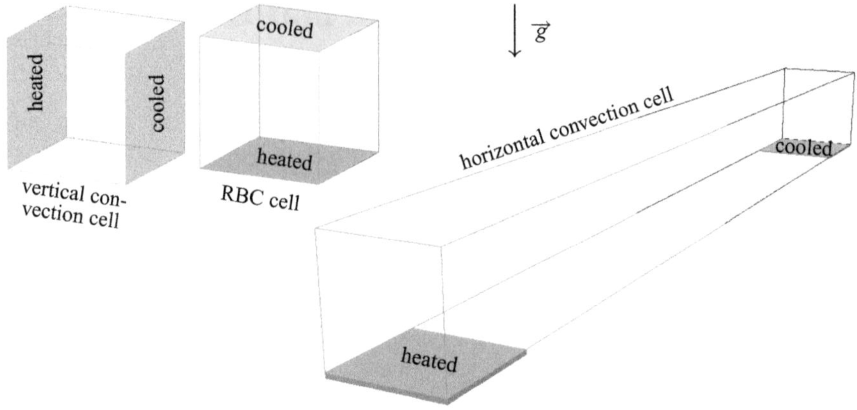

Fig. 1 Sketches of a vertical convection cell, Rayleigh–Bénard convection (RBC) cell and *horizontal* convection cell. The direction of the acceleration due to gravity g is shown with the *arrow*

The mean characteristics of the thermally driven flows are determined by the Rayleigh number $Ra \equiv \alpha g \Delta L^3/(\kappa \nu)$, the Prandtl number $Pr \equiv \nu/\kappa$, and the cell geometry. Here ν denotes the kinematic viscosity, κ the thermal diffusivity, α the isobaric thermal expansion coefficient of the fluid, g the acceleration due to gravity and $\Delta \equiv T_+ - T_- > 0$ with T_+ and T_- the temperature of, respectively, the heated and cooled plates. In RBC the reference distance L equals the distance between the heated/cooled plates, while in horizontal convection it is equal to the length of the cell, and in vertical convection it equals the width of the plates. How the mean convective heat and momentum transport, which is measured, respectively, by the Nusselt number (Nu) and Reynolds number (Re), scale with Ra and Pr is one of the main issues in studies of thermally driven flows.

2 Inclined and Vertical Convection

A principle difference between the setups in RBC and in vertical convection is in the direction of the gravity vector, which is parallel to the isothermal surfaces of the container in the case of vertical convection and is perpendicular to them in RBC. This leads to different global flow structures and to different dependences on Ra and Pr of the Reynolds number Re and mean heat flux, described by the Nusselt number Nu, in the cases of RBC and vertical convection [8, 10, 11].

Measurements in long cylinders filled with low-Prandtl-number fluids show that the convective heat transfer between the heated and cooled parallel surfaces of the container is most effective neither in a standing position of the cylinder (as in RBC, with a cell inclination angle $\beta = 0$), nor in a lying position (as in vertical convection, $\beta = 0.5\pi$), but in an inclined container, for a certain intermediate value of the

Fig. 2 Isosurfaces of the instantaneous temperature for 20 values, equidistantly distributed between the temperature of the cooled plate and that of the heated plate, in inclined convection for $Ra = 10^9$, $Pr = 1$ and the aspect ratio 1 of a cylindrical container. The case $\beta = 0$ corresponds to Rayleigh–Bénard convection, the case $\beta = 0.5\pi$ to vertical convection

inclination angle $\beta, 0 < \beta < 0.5\pi$, see, e.g., [12]. Moreover, these experiments show that in the case of small Pr ($Pr \ll 1$) and relatively large Ra ($Ra > 10^9$), any tilt $\beta, 0 < \beta \leq \pi/2$, of the cell leads to an increase of Nu, compared to that in the RBC case ($\beta = 0$).

For a fixed Ra, Pr and geometry of the container, the global flow structure depends strongly on the inclination angle β (see Fig. 2 for $Ra = 10^9$, $Pr = 1$ and also [13] for other Ra and Pr). Thus, for $Ra = 10^9$ and $Pr = 1$, in RBC ($\beta = 0$) the flow is turbulent, while in the case of vertical convection ($\beta = \pi/2$) for the same Ra and Pr one obtains almost steady flow, with hardly visible oscillations in the corners near the isothermal plates (Fig. 2).

In contrast to RBC, in vertical convection there are no exact formulas for the time and volume averaged kinetic dissipation rate ε_u and thermal dissipation rate ε_θ that relate these quantities with Ra, Nu and Pr [9]. This restricts the applicability of the Grossmann and Lohse theory [10, 11] in the case of vertical convection. On other hand, in vertical convection, in contrast to RBC, the momentum boundary layer equation for the velocity component along the isothermal plate involves the buoyancy term [8, 14]. This allows to derive the scalings of Nu and Re with Ra and Pr, at least for laminar boundary layers and for limiting cases of $Pr \to \infty$ and $Pr \to 0$. Thus, one derives for laminar vertical convection [8]:

$$Nu \sim Pr^{1/4} Ra^{1/4}, \quad Re \sim Pr^{-1/2} Ra^{1/2} \quad \text{for } Pr \ll 1,$$
$$Nu \sim Pr^{0} Ra^{1/4}, \quad Re \sim Pr^{-1} Ra^{1/2} \quad \text{for } Pr \gg 1.$$

In [8] these theoretical results were found to be in excellent agreement with direct numerical simulations (DNS) for Ra from 10^5 to 10^{10} and Pr from 10^{-2} to 30. The transition between the regimes was found to be at Pr around 10^{-1}.

Preliminary DNS for transitional and turbulent vertical convection for higher Ra show a larger exponent Υ in the scaling $Nu \sim Ra^{\Upsilon}$, i.e. $\Upsilon > 1/4$. In these turbulent regimes the Nusselt number is independent of Pr for large Pr and grows with increasing Pr as $Nu \sim Pr^{\Upsilon}$ for small Pr.

3 Horizontal Convection

For horizontal convection systems, in [5] a theoretical model for the heat and momentum transport scalings with Rayleigh number was suggested, which is an extension of the Grossmann and Lohse theory [10, 11] to the case of horizontal convection. A full diagram of the scaling regimes for the Nusselt- and Reynolds number scalings with Ra and Pr is given in [6]. The scaling regimes include in particular the Rossby scaling [15] (with the scaling exponent 1/5 in the Nu-vs. Ra scaling) and the ultimate scaling [16] (with the exponent 1/3).

The balance of the boundary-layer and bulk contributions to the time- and volume averaged kinetic and thermal dissipations rates was analysed in [17]. In the theory [5] it is assumed that within the turbulent regimes in horizontal convection, the bulk contributions should dominate. Although in the cases, studied in [17], the absolute boundary-layer contributions to the mean dissipations rates were found to prevail the bulk contributions, it was shown in [17] that the boundary-layer contributions gradually decrease with increasing Ra as soon as the horizontal convection flows become fluctuating. Further, in [18] the existence of the regimes I and I^* and the scalings within these regimes was proved by means of direct numerical simulations.

With respect the applicability of the developed theory to the global ocean circulation one can comment the following. From the oceanography [16, 19] it is known that in the planetary-scale ocean circulation there exists an equator-to-pole heat flux F of about

$$F \equiv \frac{S\rho c_p \kappa \Delta}{L} Nu \approx 2 \times 10^{15}\,\mathrm{W}.$$

The representative geometrical characteristics and fluid properties of water in the ocean can be roughly estimated as follows: the area $S \sim L^2/2$, the length $L \sim 10^7$ m, the temperature drop $\Delta \sim 10$ K, the acceleration due to gravity $g \sim 10$ m/s^2, the density $\rho \sim 10^3$ Kg/m^3, the specific heat capacity $c_p \sim 4 \times 10^3$ J/(Kg K), the thermal diffusivity $\kappa \sim 10^{-7}$ m^2/s and the kinematic viscosity $\nu \sim 10^{-6}$ m^2/s. In terms of the dimensionless quantities, this heat flux is equivalent to the Nusselt number of order 10^8, while the Rayleigh number is about 10^{31}. Note, that the Rossby scaling, if extended to $Ra \sim 10^{31}$, gives an estimate of the Nusselt number, which is at most of order 10^6 (for $Pr \approx 10$), and therefore, it cannot explain the 2 petawatts global horizontal heat flux in the ocean.

Using the theory [5, 6], one can give a possible explanation of the huge equator-to-pole heat flux. Indeed, for sufficiently large Pr and moderate Ra, the Nusselt number should grow as $Nu \sim Ra^{1/4}$. For $Pr \approx 10$, which is a rough estimate of the Prandtl number in the ocean, one can take an estimate $Nu \approx 0.14\,Ra^{1/4}$, which is in good agreement with the simulation results. The scaling $Nu \sim Ra^{1/4}$ should smoothly change to the ultimate scaling $Nu \sim Ra^{1/3}$ as soon as the boundary layers become turbulent. The latter will happen when the shear Reynolds number will exceed a critical value of $Re_s \approx 400$ [20], which corresponds to a certain critical Rayleigh number Ra_{cr}. Since $Re_s \sim Re^{1/2}$ and the mean kinetic dissipation rate in the ultimate regime is estimated as $(\nu^3/L^4)Re^3$ and as $(\nu^3/L^4)RaPr^{-2}$ [5], the critical Rayleigh number for the transition to the ultimate regime in horizontal convection for $Pr \approx 10$ can be then estimated as $Ra_{cr} \sim 4 \times 10^{17}$. Assuming that at the critical Rayleigh number, the scaling with the exponent 1/4 matches the scaling with the exponent 1/3, one can estimate the unknown prefactor in the ultimate scaling as 0.005. The resulting formula $Nu \sim 0.005Ra^{1/3}$ leads to the Nusselt numbers of order 10^8 at the Rayleigh numbers of order 10^{31}. Thus, theoretically, horizontal convection due to surface heating/cooling in the ocean can be responsible for the 2 petawatts equator-to-pole heat flux in the ocean.

Acknowledgements OS is grateful to G. Ahlers, E. Bodenschatz, E. Ching, S. Grossmann, X. He, S. Horn, D. Lohse and S. Weiss for useful discussions and acknowledges the financial support of the Deutsche Forschungsgemeinschaft (DFG) under Grant Sh405/4 – Heisenberg fellowship.

References

1. G. Ahlers, S. Grossmann, D. Lohse, Heat transfer and large scale dynamics in turbulent Rayleigh–Bénard convection. Rev. Mod. Phys. **81**, 503–537 (2009)
2. F. Chillà, J. Schumacher, New perspectives in turbulent Rayleigh–Bénard convection. Eur. Phys. J. E **35**, 58 (2012)
3. G.O. Hughes, R.W. Griffiths, Horizontal convection. Ann. Rev. Fluid Mech. **40**, 185–208 (2008)
4. R.W. Griffiths, G.O. Hughes, B. Gayen, Horizontal convection dynamics: insights from transient adjustment. J. Fluid Mech. **726**, 559–595 (2013)
5. O. Shishkina, S. Grossmann, D. Lohse, Heat and momentum transport scalings in horizontal convection. Geophys. Res. Lett. **43**, 1219–1225 (2016)
6. O. Shishkina, S. Grossmann, D. Lohse, Prandtl-number dependences of the heat and momentum transport in horizontal convection. Proc. IUTAM **00**, 000–000 (2016)
7. S. Ostrach, An analysis of laminar free-convection flow and heat transfer about a flat plate parallel to the direction of the generating body force. NACA report, vol. 1111 (1953)
8. O. Shishkina, Momentum and heat transport scalings in laminar vertical convection. Phys. Rev. E (R) **93**, 051102 (2016)
9. C.S. Ng, A. Ooi, D. Lohse, D. Chung, Vertical natural convection: application of the unifying theory of thermal convection. J. Fluid Mech. **764**, 349–361 (2015)
10. S. Grossmann, D. Lohse, Scaling in thermal convection: a unifying theory. J. Fluid Mech. **407**, 27–56 (2000)
11. S. Grossmann, D. Lohse, Thermal convection for large Prandtl numbers. Phys. Rev. Lett. **86**, 3316–3319 (2001)
12. P. Frick, R. Khalilov, I. Kolesnichenko, A. Mamykin, V. Pakholkov, A. Pavlinov, S.A. Rogozhkin, Turbulent convective heat transfer in a long cylinder with liquid sodium. Europhys. Lett. **109**, 14002 (2015)
13. O. Shishkina, S. Horn, Thermal convection in inclined cylindrical containers. J. Fluid Mech. **790**, R3 (2016)
14. O. Shishkina, S. Horn, S. Wagner, E.S.C. Ching, Thermal boundary layer equation for turbulent Rayleigh–Bénard convection. Phys. Rev. Lett. **114**, 114302 (2015)
15. H.T. Rossby, On thermal convection driven by non-uniform heating from below: an experimental study. Deep Sea Res. **12**, 9–16 (1965)
16. J.H. Siggers, R.R. Kerswell, N.J. Balmforth, Bounds on horizontal convection. J. Fluid Mech. **517**, 55–70 (2004)
17. O. Shishkina, Mean flow structure in horizontal convection. J. Fluid Mech. **812**, 525–540 (2017)
18. O. Shishkina, S. Wagner, Prandtl-number dependence of heat transport in laminar horizontal convection. Phys. Rev. Lett. **116**, 024302 (2016)
19. W. Munk, C. Wunsch, Abyssal recipes II: energetics of tidal and wind mixing. Deep-Sea Res. **45**, 1977–2010 (1998)
20. L.D. Landau, E.M. Lifshitz, *Fluid Mechanics. Course of Theoretical Physics*, vol. 6, 2nd edn. (Butterworth-Heinemann, Oxford, 1987)

Printed by Printforce, the Netherlands